# STEALTH ANTENNAS
## FOR HAM RADIO

Published by
ARRL The National Association for Amateur Radio®
225 Main Street, Newington, CT 06111-1400 USA
www.arrl.org

Printed in the USA



ISBN: 978-1-62595-228-8

First Edition, First Printing

We strive to produce books without errors. Sometimes mistakes do occur, however. When we become aware of problems in our books (other than obvious typographical errors), we post corrections on the ARRL website.

If you think you have found an error, please check **arrl.org/product-notes** for corrections. If you don't find a correction there, please let us know by sending an email to **pubsfdbk@arrl.org**.

## About ARRL

We're the American Radio Relay League, Inc. — better known as ARRL. We're the largest membership association for the amateur radio hobby and service in the US. For over 100 years, we have been the primary source of information about amateur radio, offering a variety of benefits and services to our members, as well as the larger amateur radio community. We publish books on amateur radio, as well as four magazines covering a variety of radio communication interests. In addition, we provide technical advice and assistance to amateur radio enthusiasts, support several education programs, and sponsor a variety of operating events.

One of the primary benefits we offer to the ham radio community is in representing the interests of amateur radio operators before federal regulatory bodies advocating for meaningful access to the radio spectrum. ARRL also serves as the international secretariat of the International Amateur Radio Union, which performs a similar role internationally, advocating for amateur radio interests before the International Telecommunication Union and the World Radiocommunication Conference.

Today, we proudly serve nearly 151,000 members, both in the US and internationally, through our national headquarters and flagship amateur radio station, W1AW, in Newington, Connecticut. Every year we welcome thousands of new licensees to our membership, and we hope you will join us. Let us be a part of your amateur radio journey. Visit www.arrl.org/join for more information.

**ARRL** The National Association for Amateur Radio®

225 Main Street
Newington, CT 06111-1400 USA
860-594-0200
860-594-0259
membership@arrl.org

**www.arrl.org**

# CONTENTS

## INTRODUCTION



## CHAPTER 1

### Flagpole and Vertical Antennas



## CHAPTER 2

### Loop and Slot Antennas

## CHAPTER 3

### Disguised and Novelty Antennas



## CHAPTER 4

### Low Profile Wire Antennas

# INTRODUCTION

## A Guide to the Art of Stealth

# A Guide to the Art of Stealth

Steve Ford, WB8IMY

It is common knowledge that most amateurs consider antennas to be works of art. We gaze in awestruck admiration at behemoths such as the Very Large Array in New Mexico. Even ordinary broadcast antennas can instantly arrest our vision with their soaring towers.

The Very Large Array is 50 miles west of Socorro, New Mexico. [National Radio Astronomy Observatory photo]

Since the earliest days of radio, amateurs have yearned to erect their own monuments to the magic of wireless. Back in the day, a ham may have started with a simple wire antenna in their backyard, but the ultimate goal was often a tower—the taller the better—with a Yagi and rotator at its summit. If a tower wasn't possible, a forest of wires and vertical monopoles would also do.

However, times have changed.

Large antenna systems have become prohibitively expensive for many. Not only that, but in recent decades homeowner associations (HOAs) have come to dominate the landscape. These associations exist largely to create attractive living environments where property values are protected by contracts that place restrictions on what residents can and can't do with their homes.

If you are a homeowner in an association community, you must agree to abide by the HOA rules. For example, the rules may prohibit you from painting your roof pink because its appearance would clash starkly with all the other homes and would potentially dissuade prospective buyers. Fewer buyers equal depressed property values. Everyone loses.

Unfortunately, HOA restrictions often apply to amateur radio antennas as well. After all, while we may appreciate the beauty of antennas, we must concede that this is a minority opinion. Installing a 100-ft tower and a stacked set of Yagis in your backyard may be a magnificent addition from your point of view, but your neighbors will likely regard it as a hideous monstrosity. They may even feel the same about a wire dipole antenna.

For several years ARRL, the National Association for Amateur Radio, has worked to oppose the most onerous of these HOA restrictions, especially those that completely prohibit all amateur radio activity. This work continues to this day but is a rather uphill battle against a powerful opposition.

Of course, you'll often hear people argue that if you don't want to live in antenna-restricted areas, you shouldn't buy a home in an association. This is a valid point, but today it has become increasingly difficult to find desirable properties outside of associations. And for many people—hams included—the benefits of association living far outweigh any restrictions.

And let's not forget the large portion of the population for whom home ownership is an unattainable dream. These individuals reside in rental properties such as apartments, condominiums, and rental homes. Hams living in these situations are entirely at the mercies of their landlords when it comes to antennas.

For amateurs who must live with antenna restrictions, whether imposed by landlords or HOAs, their only options are to either indulge in mobile or portable operating away from home, establish (or subscribe to) a remote station they can access through the internet, or resort to stealth.

## The Ethics of Stealth

There is no getting around the fact that stealth involves deception, but from an ethical standpoint, it also matters if the intent of the deception is malicious. For example, a stage magician may entertain you by using their sleight-of-hand skills to make it appear as though an object has suddenly vanished. They intentionally deceive you, but there is no malice involved; far from it!

However, if you deceive someone to gain an unfair advantage, or to cause harm of any sort, that is malicious. By the same token, engaging in deception to violate agreements that you've entered willingly, such as rental contracts or association agreements, is also malicious. This book in no way condones such behavior.

Ethical stealth operating simply acknowledges the idea that it is possible to enjoy amateur radio while still conforming to the rules that govern your living environment. This is why you must carefully study your rental contract or association agreement before attempting to deploy any of the antennas described in this book.

If you're a renter, keep in mind that you do not own your residence and cannot modify it in any way without the permission of the landlord. For instance, you can't drill an antenna feed line hole through a wall without permission. Doing so without first receiving approval would be effectively engaging in property damage!

If you own a home in an association, you enjoy more flexibility, but there are still limits. You might be able to hang a gaudy chandelier in your living room and paint the walls purple, but the rules may specify that your living room window curtains must be white to conform with the overall visual appearance of the neighborhood. (No kidding; such rules actually exist in some associations.)

It's best not to adopt the "what is not specifically prohibited is permitted" attitude. Most contracts and agreements are written in language to include blanket provisions that can be used to enforce unspecified restrictions. Besides, the last thing you want is to find yourself at war with a landlord or an association. This almost always ends badly.

The best approach to take is that if you've read the contract but are still in doubt, ask for permission rather than forgiveness.

## Does Stealth Mean "Invisible?"

According to the Merriam-Webster dictionary, stealth is defined as, "A cautious, unobtrusive, and secretive way of moving or proceeding intended to avoid detection or attract attention." So, while being invisible is a desirable quality for a stealthy antenna, it isn't strictly necessary.

In fact, many amateurs use antennas that are plainly visible yet unobtrusive. Consider a ham who lives in a fourth-floor apartment with a balcony. They could set up a so-called Hamstick dipole antenna on the balcony and operate to their heart's content. The antenna could be a permanent fixture on the balcony, like a chair or table, or it could be set up only when they wish to get on the air. This type of antenna isn't terribly efficient, but the balcony's height advantage may compensate considerably. Best of all, it is unlikely to attract anyone's attention.

**This single-band HF dipole antenna consists of two Hamstick-style mobile antennas joined together. The antenna can be set up, and taken down, in a matter of minutes. [WB8IMY photo]**

Or consider an amateur who lives in a development with an HOA that forbids permanent outdoor antenna installations. When they want to get on the air, they might set up a compact HF loop antenna on their backyard deck (or on the lawn) and remove it when they're finished. No doubt some neighbors may notice the loop, but they will also notice that it's only present for a limited time. No harm, no foul.

If you prefer an antenna that is invisible, camouflaged, or otherwise truly stealthy, the design you'll choose will depend on key considerations including the structure of your dwelling, the availability of supports, such as trees or flagpoles, and more importantly the

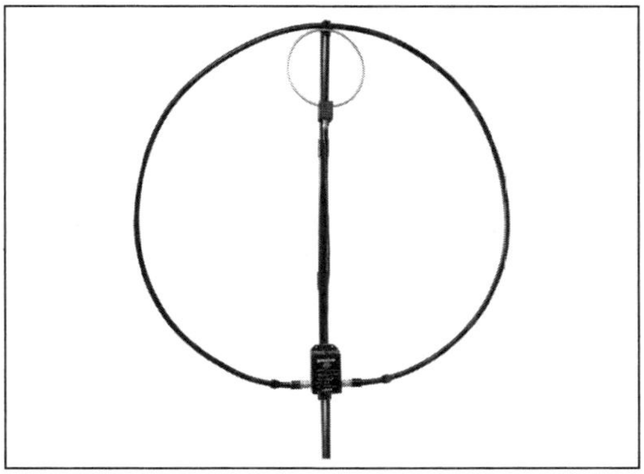

**A compact HF magnetic loop antenna can become a temporary solution in a restricted-antenna environment. [WB8IMY photo]**

wording of your rental/lease contract or association agreement.

Aesthetics may be an additional concern when selecting an antenna. Let's say you find yourself in one of those uncommon situations where your landlord or association is tolerant of your desire to set up an outdoor antenna. You may be able to sweeten the deal by pledging to install an antenna that has the lowest visual profile possible, a consideration that your neighbors will likely appreciate.

And finally, when selecting a stealth antenna, there is always the issue of safety. For instance, radial wires on lawns present serious tripping hazards, as do feed lines. Yes, you can bury both beneath the soil, but only if you are allowed to do so. Never take it upon yourself to start digging up the lawn!

Another safety concern applies to the RF energy you'll be radiating from your antenna. Anyone coming into contact with your antenna while you are transmitting is likely to receive an unpleasant shock, depending on how much power you are running, so careful placement to avoid such encounters must be considered. There is also the issue of RF exposure, which we'll discuss in a later section.

## Making the Most of Compromises

All stealth antennas, including the ones described in this book, are "compromises." This means that while each antenna is designed to achieve one or more goals, most will not offer the same performance as full-sized antennas.

A stealth antenna is typically designed to present a reasonably good impedance match so that the transceiver can load its output power to the system with the greatest efficiency possible. But even with the best design intentions, efficiency may suffer; it is just a matter of degree.

For example, HF mobile antennas are notoriously inefficient, losing a significant amount of RF energy as heat. However, these antennas are also physically short, which makes it possible to use them on various vehicles—and as stealth antennas. They effectively trade radiation efficiency for greatly reduced size compared to, say, a 1/4-wavelength design. In other words, an HF mobile antenna represents a compromise.

As a "stealthy" operator, you need to keep these compromises in mind and be prepared to accept reduced performance because of them. This is not to say that you will be in abject misery every time you turn on your radio, but it does mean that you'll be a much happier ham if you know what to expect and operate accordingly. For example, if you attempt to answer a CQ call from a weak station, don't be surprised if you are unsuccessful. Chances are that your signal is just as weak to the other operator. You are likely to enjoy greater success if you seek the stronger stations. Thanks to the concept of signal reciprocity, if you are receiving them well, they are likely to hear you, too.

Stealth operating can be highly rewarding if you understand your limitations and work to alleviate them as much as possible. One solution is to increase your output power. You'll consistently enjoy more success if you are running 50 or 100 W output to your antenna rather than 5 W. Boosting your output is often problematic, though. You are likely to cause more interference to your neighbors (and yourself), and higher power levels also raise the specter of RF exposure. We'll discuss these problems in more detail shortly.

If you can't increase your RF output, what is the alternative? Changing your operating mode!

### Pick the Best Mode for the Job

One important characteristic of any operating mode is the bandwidth its signal occupies. SSB spreads your transmitted energy over a little less than 3 kHz of RF spectrum. In contrast, a CW signal may occupy only 50 Hz.

Why does signal bandwidth matter? Imagine a garden hose with an adjustable nozzle. If you set the nozzle for the "fine spray" mode, the water is diffused into a wide cone of droplets. The droplets will strike the ground at your feet, and the ground will certainly become wet, but each square inch of soil will receive relatively little water overall as the droplets will land with little energy. However, if you switch the nozzle to the "jet" mode, all the water in the hose is suddenly focused into a narrow stream. The water strikes the ground with considerable force, perhaps enough to start ripping away the soil. A high-pressure cleaner works the same way.

With the water analogy in mind, generating an SSB signal is like setting the nozzle to "fine spray." You're spreading your precious RF energy over a wide slice of spectrum and the receiving station will have to capture a fair amount of it to understand what you are saying.

But if you switch to a mode such as CW, you're effectively setting the nozzle to "jet" and your RF energy is concentrated within a much narrower bandwidth. The receiving station doesn't have to capture as much of your signal to decode what you are sending.

Hams with full-size antennas can compensate for the diffused energy of an SSB signal by using directional designs that act like magnifying glasses to focus RF energy. They can also transmit at high power levels, up to 1.5 kW, depending on the frequencies and their license privileges. High RF power and a directional antenna is a potent combination!

But when you are restricted to low power (less than 100 W) and compromise antennas, you need to compensate by using modes that either concentrate your RF energy in a narrower bandwidth, use digital encoding and processing, or both. This is not to say that you can't operate SSB with a stealth antenna system, just that the chances are good that you'll find it somewhat frustrating. On the other hand, you may discover that you can routinely make contacts throughout North America with just 5 W of CW applied to some highly inefficient antennas. This type of low-power operating, known as QRP, is extremely popular among stealth operators. There is nothing magical about QRP. In truth, your success with QRP depends in large part on the other operator's station and their ability to pull your signal out of the noise. The good news is there are plenty of higher profile stations available to contact. The other great thing about QRP is that it usually doesn't wreak havoc with your neighbor's consumer electronic devices.

If CW isn't your game, fear not; you have alternatives in the digital realm that are every bit as effective for stealth operating as CW, if not more so. A basic HF digital station consists of little more than an SSB transceiver, a

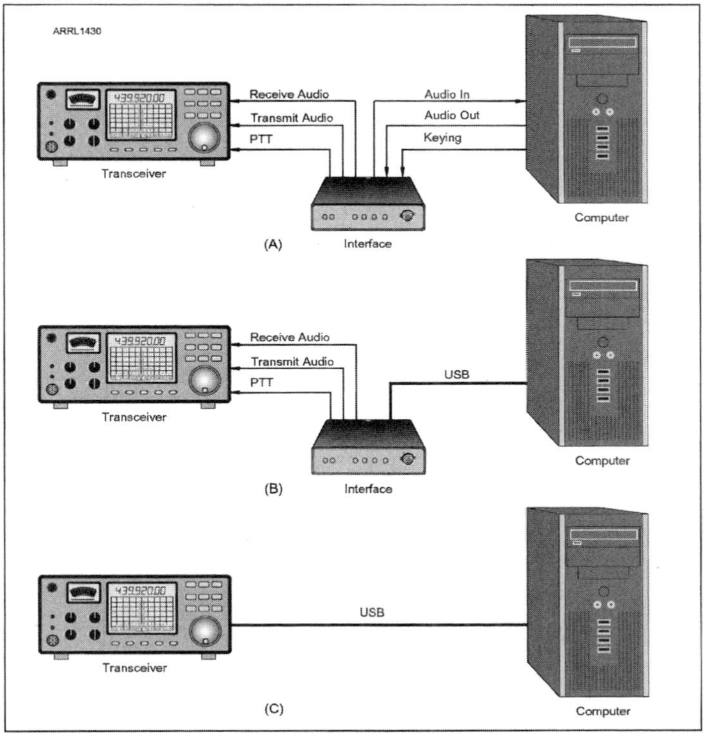


**Figure 1** – Three of the most common configurations for interfacing your transceiver with your computer for stealthy digital operating. (A) The most common setup for many years was based on an interface that took transmit and receive audio from the computer sound card, along with transmit/receive keying from one of the computer's COM ports and passed everything to the transceiver in a way that kept the signal lines isolated. This type of interface is increasingly uncommon. (B) Many interfaces available at the time this book was written are USB devices. A single USB cable plugs into the computer. Not only is the computer sound card not used, but the interface also contains its own sound device. In turn, the interface supplies transmit audio to the radio, processes receive audio from the radio and handles transmit/receive keying. (C) The future will likely see the disappearance of the interface entirely as transceivers incorporate their own computer interfacing. Today, a number of rigs have added this feature.

computer, and an interface device (see **Figure 1**). Depending on the type of transceiver you own, it may already have the computer interface built in. In that case, all you need is the computer.

A complete explanation of HF digital operating is way beyond the scope of this article because there are so many different modes in use (about 60 at last count). For instance, there is radio teletype or RTTY, an old digital mode that is still popular for contesting and DXing. There are also advanced error-free chat applications such as VarAC (see **varac-hamradio.com**).

And then there is FT8, possibly the most popular mode among stealth operators. With FT8 you will often be able to work a ton of DX stations, even with very low power. The downside of FT8 is that you won't be able to carry on conversations with these stations. FT8 exchanges only the minimum information necessary for a valid contact, one that counts for awards such as the ARRL DX Century Club. If want to have a true conversation, you will need to explore modes such as VarAC, or an FT8 variant known as JS8.

For a detailed overview of digital operating, pick up a copy of ARRL's *Get On the Air with HF Digital*. You can find it at the ARRL online store (**arrl.org/shop**) as well as other online retailers such as Amazon.

## Coping with Noise and Interference

As a stealth operator, it is likely that your antenna will be quite close to you and your neighbors. With this being the case, there are two types of radio frequency interference (RFI) that you may encounter: (1) Interference you

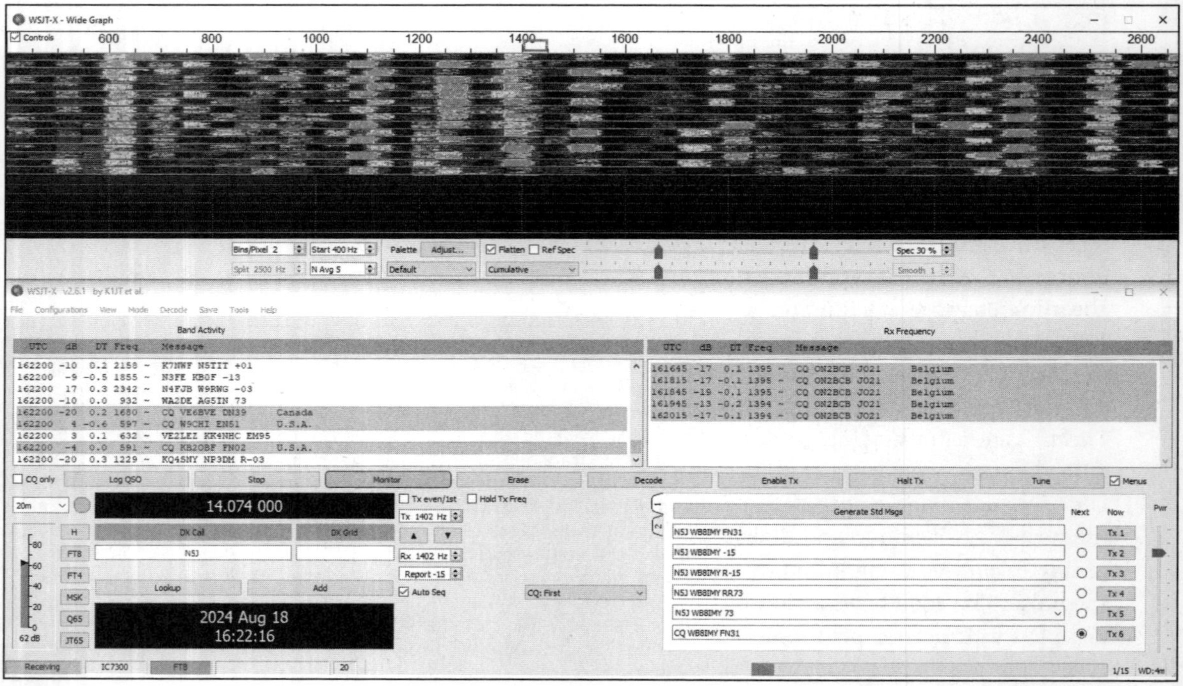

**FT8 is one of the most popular modes among stealth operators. [WB8IMY photo]**

inflict upon yourself and your neighbors and (2) interference that your consumer electronic devices, and those of your neighbors, inflict upon you.

### When You're the Cause: Turn Down the Power

If your transmissions are causing interference, the easy first step is to reduce your output power. As we discussed earlier, low-power operation is best for the stealth ham. At the very least, if you are attempting to operate at the 100 W output level, consider reducing your output to 50 W. You'll notice little change in your ability to make contacts, but you may see a major change in the severity of the interference you are causing.

In most instances, reducing your output power is a quick cure. Simply reduce the power in small steps, pausing to see if you are still generating interference. Chances are that you won't need to make a large reduction to cure the problem. If you operate on several bands, you may also discover that your station causes fewer problems on some bands than others.

### When You're the Victim: Identifying the Noise Source

It is useful to place an offending noise source into one of several broad categories at the initial stages of any RFI investigation. Since locating and resolution techniques can vary for each type, the process of locating and resolving RFI problems should begin with identifying the general type of RFI source.

It is often impossible to identify the exact type of device generating the RFI from the sound of the interference. Because there are many potential sources of RFI, it is often more important to obtain and interpret clues from the general noise characteristics and the patterns in which it appears.

A source that exhibits a repeatable pattern over the course of a day or week, for example, suggests something associated with human activity. A sound that varies with or is affected by weather suggests an outdoor source. Noise that occurs in a regular and repeating pattern of peaks and nulls as you tune across the spectrum, every 50 kHz for example, is often associated with switchmode power supplies or similar pulsed-current devices. A source that exhibits fading or other sky wave characteristics suggests something that is not local. A good ear and careful attention to detail will often turn up some important clues. A detailed RFI log can also help, especially if maintained over time.

Noise can be characterized as broadband or narrowband — another important clue. Broadband noise is defined as noise having a bandwidth much greater than the affected receiver's operating bandwidth and is uniform across a wide frequency range. Noise from arcs and sparks, such as power line noise, tends to be broadband. Narrowband noise is defined as noise having a bandwidth less than the affected receiver's bandwidth. Narrowband noise is present on specific, discrete frequencies or groups of frequencies, with or without additional modulation. In other words, if you listened to the noise on an SSB receiver, tuning would cause its sound to vary, just like a regular signal. Narrowband noise often sounds like an unmodulated carrier with a frequency that may drift or suddenly change. Microprocessor clock harmonics, oscillators, and transmitter harmonics are all examples of narrowband noise.

### Identifying Noise from Part 15 Devices

Part 15 devices and other consumer equipment noise sources are ubiquitous. Although the absolute signal level from an individual noise source may be small, their increasing numbers makes this type of noise a serious problem for stealth amateur radio stations.

Electronic devices containing oscillators, microprocessors, or digital circuitry produce RF signals as a byproduct of their operation. The RF noise they produce may be radiated from internal wiring because of poor shielding. The noise may also be conducted to external, unshielded, or improperly shielded wiring as a common-mode signal where it radiates noise. Noise from these devices is usually narrowband that changes characteristics (frequency, modulation, on-off pattern) as the device is used in different ways.

Another major class of noise source is equipment or systems that control or switch large currents. Among them are variable-speed motors in products as diverse as washing machines, elevators, and heating and cooling systems. Charging regulators and control circuitry for battery and solar power systems are a prolific source of RF noise. So are switchmode power supplies for computers and low-voltage lighting. This type of noise is only present when the equipment is operating.

Switchmode supplies, solar controllers, and inverters often produce noise signals every N kHz, with N typically being from 5 to 50 or more kHz. This is different from noise produced by spark or arc sources that is uniform across a wide bandwidth. This pattern is often an important clue in distinguishing switching noise from powerline or electrical noise.

Wired computer networks radiate noise directly from their unshielded circuitry and from network and power supply cables. The noise takes two forms: broadband noise and modulated carriers at multiple frequencies within the amateur bands. As an example, Ethernet network interfaces often radiate signals heard on a receiver in CW mode. Each network interface uses its own clock, so if you have neighbors with networks, you'll hear a cluster of carriers around these frequencies, ±500 Hz or so.

In cable TV systems, video signals are converted to RF across a wide spectrum and distributed by coaxial cable into the home. Some cable channels overlap with amateur bands, but the signals should be confined within the cable system. No system is perfect, and it is common for a defective coax connection to allow leakage to and from the cable. When this happens, a receiver outside the cable will hear RF from the cable and the TV receiver may experience interference from local transmissions.

### Hunting Down RFI

Unless you live in an apartment or condominium, most RFI sources will be found inside your own home. Locating an in-house source of RFI is so simple that it makes sense to start an investigation by simply turning off your home's main circuit breaker while listening to the noise with a battery-powered portable radio. (Don't forget that battery-powered equipment may also be a noise source, so remove the batteries from consumer devices as well.) If the noise goes away, you know the source is in your residence. After resetting the breaker, you can further isolate the source by turning off individual breakers one at a time. Once you know the source circuit, you can then unplug devices on that circuit to find it.

**Caution:** Do not attempt to remove cartridge fuses or operate exposed or open-type disconnects if it is possible to make physical contact with exposed electrical circuits. Also, if you live in an apartment or condominium, make sure the circuits you are turning off are within your residence only.

It is often possible to locate a noise source

outside your home with a minimum of equipment and effort. Because of Part 15's absolute emissions limits, most Part 15 noise sources are within a few hundred feet of your antenna.

Electrical noise sources in a home, such as an arcing thermostat or a noisy washing machine controller, can also be tracked down in the same way as noise from consumer electronic devices. Electrical noise from an incidental emitter, such as a power line, can propagate much farther than noise from an otherwise legal unintentional emitter. Some Part 15 devices, for example battery chargers for electric scooters and wheelchairs, are notorious for exceeding Part 15 absolute emissions limits on conducted noise.

The following procedures are intended for areas with detached houses rather than apartment or condominium units. However, you can adapt them accordingly.

**One word of warning:** We live in a time when people are particularly wary of the activities of strangers, especially strangers snooping around with radios. If your neighbors see you acting in a way that they think is suspect, there is a strong probability that you'll soon find yourself face to face with a police officer. While hunting interference outside your home, always consider your own safety and the safety of those around you. Dave Cole, NK7Z, a frequent RFI hunter, divides the world of RFI exploration outside his home into two separate categories: people he knows and people he doesn't. If Dave knows his neighbor well enough to have a conversation with him or her, he is likely to approach them about RFI problems emanating from their homes. But if he doesn't know the neighbor well, he will reach out to the ARRL lab to start the process of official communications from the FCC to address the problem.

Never walk onto someone's property and begin probing around their residence without permission—even if you strongly suspect that the interference culprit is inside their home. Once again, this approach is likely to result in a 911 call, or worse!

As you stroll casually along the street or sidewalk, use a detector suitable to receive the noise, typically a battery-powered receiver. Preferably, the receiver should have a variable RF gain control. An external step attenuator will also work if the antenna is external to the radio. If the antenna can be removed, a probe can also be made from a small piece of wire or paper clip to reduce the receiver's sensitivity.

[*Ed. note* — These tools will help you figure out where the noise source is. This task is called radio direction finding (RDF), and for a more suitable in-depth treatment of antennas suitable for RDF see chapter 21 of the *ARRL Handbook for Radio Communications.*]

The importance of personal diplomacy simply cannot be overstated once you identify the source residence and make the decision to approach your neighbor. The first contact regarding an RFI problem between a ham and a neighbor is often the most important; it is the start of all future relations between the parties. The way you react and behave when you first discuss the problem can set the tone for everything that follows. It is important, therefore, to use a diplomatic path from the very start.

Greet your neighbor with a radio in hand, preferably an ordinary AM broadcast or shortwave receiver. Let them hear the noise, but not so loud that it will be offensive. Tell them this is the problem you are experiencing, and you believe the source may be in their home. Don't suggest what you think the cause is. If you're wrong, it often makes matters worse. Tell them it will only take a minute to determine whether the source is in their home. Most neighbors will agree to help find the source, and if they agree to turn off circuit breakers, it can be found very quickly. Start with the main breaker to verify you have the correct residence, then the individual breakers to find the circuit.

### Interference Fixes

Regardless of whether the RFI source is in your home or a neighbor's home, there are several steps you can take to either cure the problem completely or at least reduce the severity.

Filters and chokes can be highly effective in dealing with interference problems. Use of chokes alone can often solve many RFI issues, especially at HF when common-mode current is more likely to be the culprit.

A primary means of separating signals relies on their frequency difference. Filters offer little opposition to signals with certain frequencies while blocking or shunting others. Filters vary in attenuation characteristics, frequency characteristics, and power-handling capabilities. The names given to various filters are based on their uses:

***Low-pass filters*** pass frequencies below some cutoff frequency, while attenuating frequencies above that cutoff frequency. These filters are difficult to construct properly, so your best bet is to purchase one. Many amateur radio retail stores that advertise in *QST* magazine stock low-pass filters.

***High-pass filters*** pass frequencies above some cutoff frequency while attenuating frequencies below that cutoff frequency. Again, it is best to buy one of the commercially available filters.

***Bypass capacitors*** can be used to cure RFI problems by providing a low-impedance path for RF signals away from the affected lead or cable. A bypass capacitor is usually placed between a signal or power lead and the equipment chassis. If the bypass capacitor is attached to a shielded cable, the shield should also be connected to the chassis. Bypass capacitors for HF signals are usually 0.01 μF, while VHF bypass capacitors are usually 0.001 μF. Leads of bypass capacitors should be kept short, particularly when dealing with VHF or UHF signals.

***AC-line filters,*** sometimes called "brute-force" filters, are used to remove RF energy from ac power lines. Products from TE Connectivity, (**te.com/en/home.html**) and others are widely available.

AC-line filters come in a wide variety of sizes, current ratings, and attenuation. In general, a filter must be physically larger to handle higher currents at lower frequencies. The Corcom 1VB1, a compact filter small enough to fit in the junction box for many low voltage lighting fixtures, provides good common mode attenuation at MF and HF, and its 1 A at 250 V ac rating is sufficient for many LV lighting fixtures. In general, you will get more performance from a filter that is physically small

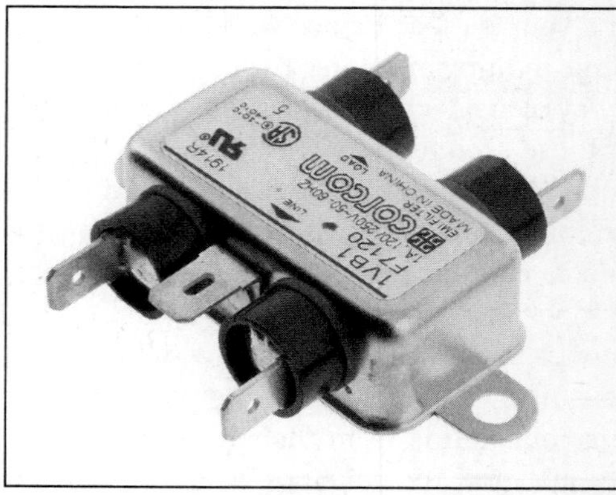


The Corcom 1VB1 ac line filter. [WB8IMY photo]

if you choose the filter with the lowest current rating sufficient for your application.

### Common-Mode Chokes

A *common-mode* signal refers to a component of a signal that is common to both input lines of a differential pair of wires. A common-mode signal is the portion of the signal that is identical on both conductors relative to a common reference point, typically ground. This signal does not contribute to the differential signal, which is the difference between the voltages on the two lines. Instead, common-mode signals can be the result of external noise, power supply fluctuations, or even imperfections in the circuit design. These signals are often the bane of amateur radio stations, causing interference and other forms of strange behavior.

Common-mode chokes on ferrite cores are among the most effective answers to RFI from a common-mode signal. Common-mode chokes work differently, but equally well, with coaxial cable and paired conductors. The most ordinary form of common-mode choke is multiple turns of cable wound on a magnetic toroid core, usually ferrite. Most of the time, a common-mode signal on a coaxial cable or a shielded, multi-wire cable is a current flowing on the outside of the cable's shield. By wrapping the cable around a magnetic core, the current creates a flux in the core, creating a high impedance in series with the outside of the shield. (An impedance of a few hundred to several thousand ohms are required for an effective choke.) The impedance then blocks or reduces the common-mode current. Because equal-and-opposite fields are coupled to the core from each of the differential-mode currents, the common-mode choke has no effect on signals inside the cable.

You can easily create an effective common mode choke by looping several turns of wire through a ferrite torroid. [WB8IMY photo]

When the cable consists of two-wire, unshielded conductors such as zip cord or twisted-pair, the equal-and-opposite differential currents each create a magnetic flux in the core. The equal-and-opposite fluxes cancel each other and the differential-mode signal experiences zero net effect. To common-mode signals, however, the choke appears as a high impedance in series with the signal: the higher the impedance, the lower the common-mode current.

It is important to note that common-mode currents on a transmission line will result in radiation of a signal from the feed line. The radiated signal can then cause RFI in nearby circuits. This is most common when using coaxial cable as a feed line to a balanced load, such as a dipole antenna. Using a common-mode choke to reduce common-mode feed line currents can reduce RFI caused by signals radiated from the feed line's shield.

The optimum core size and ferrite material is determined by the application and frequency. For example, a cord with a plug attached cannot be easily wrapped on a small ferrite core. The characteristics of ferrite materials vary with frequency. Type 31 material is a good all-purpose material for HF and low-VHF applications, especially at low HF frequencies. Type 43 is widely used for HF through VHF and UHF.

You'll often run into clamp-on split cores. These are essentially single-turn chokes since the cable passes just once through the bead or core. Multi-turn chokes are required for effective suppression at HF. It is usually more effective to form a common-mode choke by wrapping multiple turns of wire or coaxial cable around a 1- to 2-inch OD core of the correct material. Otherwise, you'd have to clamp quite a few cores on a cable that you think is causing (or being subjected to) RFI.

### Another Approach to Battling Interference

If your station is the victim of stubborn interference that no application of shielding, chokes, or appeals to your neighbors can mitigate, you may want to try an interference-cancelling device. Interference-cancelling devices (often called "noise cancelers") function in a manner similar to noise-canceling headphones. They sample the noise in the environment, invert its phase relationship, and then mix this phase-inverted signal with the original signal. The result is that the phase-inverted signal effectively cancels the noise in the original signal. It is ingenious technology!

There are devices available that can work the same magic for RF signals. One example shown here is the Timewave ANC-5. The ANC-5 works by receiving local interference signals with a short "sense" antenna, inverting the phases of these signals, and then feeding them back into your radio's primary antenna port where they cancel each other out. Devices such as the ANC-5 can have an amazing effect on interfering signals, so long as the signals are from nearby sources (such as computers, washing machines, or solar panels).

There are a couple of drawbacks to using a device such as an ANC-5. One is cost, which can typically be as high as $250 or more. The other is the fact that each time you change frequencies, or frequency bands, you will likely need to readjust the canceller's controls. Even so, many amateurs find that these interference cancellers are highly effective and well worth the investment.

**The Timewave ANC-5 noise canceller. [Timewave Technology photo]**

## RF Exposure

The question of the safety of RF energy is controversial, and science has yet to determine whether a harmful relationship definitively exists between RF energy and human health. The exceptions apply to the most extreme instances where people have been exposed to very high-power levels, usually at very high frequencies (microwave ovens come to mind).

Even so, FCC regulations set limits on the maximum permissible exposure (MPE) allowed from the operation of radio transmitters. Following these regulations, along with the use of good RF practices, will make your stealth station as safe as possible.

## A Simple Exposure Assessment

RF safety rules include distinctions for people who know about their exposure and are trained in how to control it, and those who do not. The occupational exposure (also referred to as controlled exposure) designation applies to amateurs and members of their households, with the expectation that they be trained about RF exposure. A separate general population exposure designation applies to everyone else. Their exposure is deemed uncontrolled, and they have more restrictive maximum permissible exposure (MPE) limits. Most stations can be assessed with little effort. In place of the categorical exemptions, the FCC developed a table of exemption formulas based on power to the antenna, antenna gain, and minimum distance between a person and the antenna.

Assessing the RF safety of your station is much easier than you think. ARRL provides an RF Exposure Calculator at **arrl.org/rf-exposure-calculator**. The calculator requires values to be entered for the power transmitted minus the feed-line loss, the antenna gain, the type of modulation used, the approximate transmit and receive times, and the frequency of operation. There is a checkbox to enable calculations with ground effects, and in the interest of conservatism, this should be selected. From these values, the calculator generates the minimum distance from any part of the antenna that a person in the occupational and general populations may be located.

The evaluation must be repeated for every frequency band, antenna, and the maximum power level that's used to obtain a complete environmental analysis of a station.

The FCC does not require that the results of a station's exposure analysis be submitted. However, it is advisable to keep a record of the analysis so that if there's ever an exposure complaint, the calculations can be shown to the FCC.

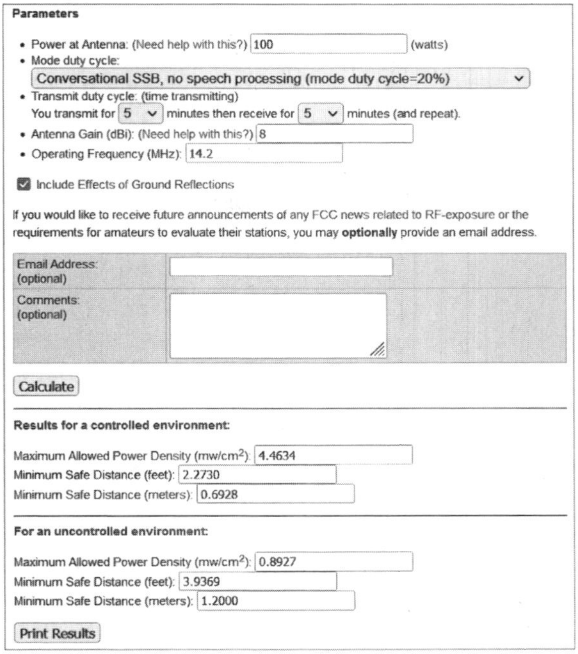
**Parameters**

- Power at Antenna: (Need help with this?) 100 (watts)
- Mode duty cycle:
  Conversational SSB, no speech processing (mode duty cycle=20%)
- Transmit duty cycle: (time transmitting)
  You transmit for 5 minutes then receive for 5 minutes (and repeat).
- Antenna Gain (dBi): (Need help with this?) 8
- Operating Frequency (MHz): 14.2

Include Effects of Ground Reflections

If you would like to receive future announcements of any FCC news related to RF-exposure or the requirements for amateurs to evaluate their stations, you may **optionally** provide an email address.

Email Address: (optional)

Comments: (optional)

Calculate

**Results for a controlled environment:**

Maximum Allowed Power Density (mw/cm$^2$): 4.4634
Minimum Safe Distance (feet): 2.2730
Minimum Safe Distance (meters): 0.6928

**For an uncontrolled environment:**

Maximum Allowed Power Density (mw/cm$^2$): 0.8927
Minimum Safe Distance (feet): 3.9369
Minimum Safe Distance (meters): 1.2000

Print Results

**Assessing the RF safety of your station is easy with the ARRL RF Exposure Calculator at arrl.org/rf-exposure-calculator.**

Try the online RF Exposure calculator; you may be pleasantly surprised at the results. Let's run the calculator for a few hypothetical situations based on stealth antennas and modes we've discussed so far. We'll use a worst-case scenario by assuming you are using a 100% duty cycle mode such as FT8 (in the calculator, we will choose FSK/RTTY). We will also assume an antenna gain of 2 dBi, although most antennas you are likely to use will have less than 2 dBi gain.

If you're operating FT8 with 100 W output on 28.074 MHz, the closest anyone in a controlled environment, such as you or a person in your home, should come to your antenna when you are transmitting is about 5 feet. For your neighbor in an adjacent apartment, the closest distance is 8 feet.

If you lower the RF output to 50 W, which is more than enough for FT8, your station will still be complying if people in your home are more than 4 feet from your antenna; any neighbors should be at least 6 feet away.

Lower the output further, and lower the frequency at the same time, and the results are even better. Let's assume a 10 W output with FT8 on 14.074 MHz. Now the minimum uncontrolled distances drop to less than 2 feet and the controlled distance is less than a foot!

Remember: you want to be prudent, not paranoid. While science has yet to establish a strong correlation between lower levels of RF energy and health risks, it is best to be conservative. On the other hand, don't allow unfounded fears to keep you from enjoying amateur radio.

Once you have your antenna installed, simply estimate the distance to the nearest uncontrolled environment for the band (or bands) on which you want to operate, and the power you wish to use, and then run the calculations. Do the same for the controlled environment (the closest anyone, including you, should come to your antenna).

Write down or print the results, then file them in a safe place in case you ever need to prove you performed the calculations. You're done!

# CHAPTER 1

# Flagpole and Vertical Antennas

# Constructing a Flagpole Antenna

*Sometimes adversity can work to a ham's advantage.*

Geoff Haines, N1GY

In the middle of the 2008 holiday season, a kind of catastrophe struck at our house. A significant amount of damage was caused by a water line that broke during the night. My wife, Audrey, and I eventually had to find a new home. This, in a strange way, turned out to be a blessing in disguise. Our new home is only about two blocks from our old one, but it is considerably newer and actually has a spare room that I immediately requisitioned for the "radio room."

## Getting Back on the Air

As soon as the domestic necessities of moving and setting up housekeeping in our new home were taken care of, work began on the location of my station. Most of the furniture from the old radio room was used, if in a slightly different layout, the radios were hooked up and power was run. Some very good friends of mine from our local club, The Manatee Amateur Radio Club, volunteered to help me get the VHF and UHF antennas mounted on my roof. Coax was run to the radio room and I was soon back on the air. The only difficulty was found to be the HF side of things.

At our previous residence, I had used a commercial multi-band HF antenna for several years. It served me well and rarely needed any attention. My mounting system was an old telescopic TV type mast with a universal joint at the base to permit the antenna and mast to be tilted to the ground when a hurricane was imminent. The mast was secured to the roof at two places and never gave us a problem.

## A Horse of a Different Hue

The new home is very different from the old place. It is oriented parallel to the road instead of at right angles to it. It is twice as wide and has virtually no backyard at all. To put the icing on the cake, the utility wires and poles run right behind the house and thus preclude any kind of a backyard setup for an antenna. The HF antenna would have to go right in the middle of the front lawn! Audrey, my wife, took one look at my old vertical and said "No." Actually, she said considerably more than that, but let's leave that conversation out of this discussion. A new plan had to be found.

I did notice that many of my neighbors had flagpoles in their front yards and so the thought occurred to me that maybe that was a solution to my dilemma. You will note that I have not mentioned anything about restrictive covenants. That is because there aren't any where I live. When we first moved to the development a number of years ago, I specifically asked the park owner about Amateur Radio antennas. His answer was simple: "As long as it doesn't look like the Johnson Space Center, you'll be okay." To that end, I have steered clear of large dish antennas and HF Yagis and everyone seems happy.

## Run it up the Flagpole

No, the problem here was one of simple aesthetics, and the judge of what would be acceptable sleeps right beside me every night. An antenna combined with a flagpole seemed to be the right way to go. At the Orlando Hamcation in February, I purchased a remote HF automatic antenna tuner. After diligent research into what would best suit my situation I chose an ICOM AH-4 based on several factors — size, compatibility with my ICOM IC-706 MkIIG, and ease of connection to the flagpole. For your station, another tuner may be more appropriate — as long as it will automatically match a wide range of loads and handle your transmitter's power.

Now I had a site (the middle of the front lawn), I had a tuner (the AH-4), and I had a plan for the radial runs under the sod. What was I forgetting? Oh yeah, the antenna/flagpole! There are several manufacturers of flagpole type antennas, and other antennas that can be disguised as a flagpole. I even visited with a fellow ham who built a flagpole around a commercial trap vertical that was featured in an issue of *QST* a few years ago.[1] I was all set to follow one of those paths when a member of our ham club mentioned that I could have an old sailboat mast from his back yard. What a find! For free, the price was right. The overall length was 22 feet. It even broke down into two sections, so it would be easy to get home. It was made of aluminum tubing, was lightweight, and even had some of the hardware attached so turning it into a flagpole would be simple.

## Making it Happen

Initially, I thought that this would be dirt simple. Then I realized that I had to mount it so that it would stay vertical. I also had to design some way to take it down should one of our frequent hurricanes head my way. Here is where the sticky bit began. The mast has a diameter of 2½ inches at the bottom. This diameter transitions to 2 inches at the junction of the two mast sections. I checked with several vendors of tilt over mounts and all were very helpful, but none had a mount that could handle anything over 2 inches of mast diameter.

And I did want a tilt over mount. I don't know about anyone else, but I did not want to try balancing a 22 foot mast while trying to lower it gently to the ground. I wanted to be able to remove one or two bolts and smoothly walk the mast down to a horizontal position. At that point I can remove the other bolts and disassemble the mast in relative comfort.

Out came the graph paper and pencils (and the erasers too, if you must know). I tried to avoid reinventing the wheel and looked closely at the various tilt mounts available commercially. From one I took the idea of a solid base plate with the antenna/flagpole attached via U bolts. I insulated the mast from the mount with a section of PVC pipe — the darker type used for electrical conduit, since the white stuff is not tolerant of too much UV light. The white PVC works fine for situations in which not too

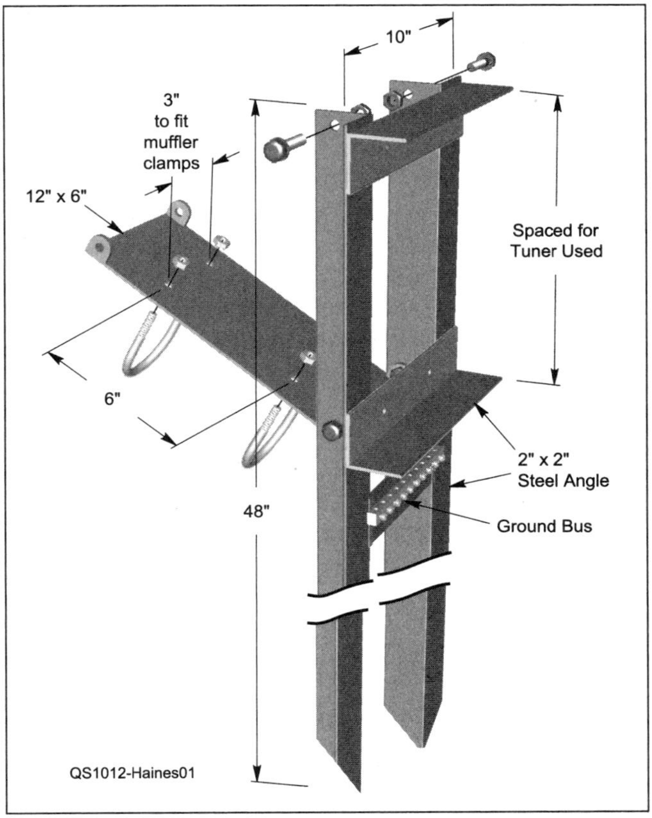


Figure 1 — The mount for the aluminum mast as fabricated. Also indicated is the final location for the AH-4 tuner. The dimensions shown apply to the author's antenna. Different tuners or different antenna masts might require a change in dimensions. Isometric view of the mount. The sideways U shapes indicate the muffler clamps that hold the mast to the mount. By removing the upper two bolts (as shown) holding the left plate in place, the antenna can be swiveled down.

much stress is placed on it, such as a radome for a VHF antenna or the like, but with even a lightweight 22 foot mast, the extra support of the slightly thicker gray PVC is appreciated.

From another manufacturer I pinched the idea of a kind of pickle fork mount, driven deep into the ground. If necessary, this, like most others, can be surrounded with a concrete footing to add weight and permanence to the mount. In evaluating the other flagpoles in our neighborhood, I found that most of them were simply supported by driving a PVC sleeve into the ground and placing the pole into the sleeve. Some, however, had been mounted in a concrete filled hole dug into the soil.

I figured that I would wait until the time came to actually place the mount before making that decision. More on that later.

**Figure 1** shows my basic design. I deliberately oversized the bolts and used steel rather than aluminum. Not being a structural engineer, I figured the safe approach was to look at what the commercially available units were made of and go up at least two notches in size. I took my design to a local welding shop and after a few modifications suggested by the owner, he proceeded to whip up the whole thing in just about one hour. The cost was no more than some of the commercial tilt mounts on the market.

After the mount was constructed, it was time for paint. I chose to paint the mast and the mount with white metal primer first, and then finish with a good grade of exterior white enamel. The finial on the very top of the mast, to be mentioned a little later, was painted with a spray can of metallic gold color paint so that the entire assembly would look as traditional as possible.

When I began to attempt driving my steel mount into our front lawn, I quickly realized that the ground under our lawn was much harder than that of my neighbors. I immediately borrowed a post hole digger from one of the neighbors and proceeded to dig a two and a half foot deep hole with it. The hole wound up sort of oval in shape because I had to allow

Figure 2 — View of the finial on top of the mast shows how a toilet tank float and a little wire can be used to extend the antenna to a full 23 feet.

for the fact that the mount had two legs. After the mount was placed in the hole, with a little gravel for drainage at the bottom, I poured in a bag of quick setting concrete obtained from the local home improvement store. One gallon of water was added as per the instructions on the bag. By this time the mount had been braced and held level and perpendicular with an assortment of scrap wood braces, to be removed once the concrete had set. With all of that done, I could now begin the process of erecting the flagpole.

I let the concrete cure for more than 24 hours before I removed the wood braces. Since I had already attached the tilting portion of the mount to the bottom section of the flagpole, it was relatively easy to attach the tilting plate to the steel uprights of the mount using the two large bolts that fit into the nuts welded on the back side of the plate. With Audrey's help, I then walked the flagpole up to the vertical plane. While she steadied it, I inserted the other two bolts at the top of the plate to hold it firmly perpendicular to the lawn. Before raising the mast to its final position, I added a 12 inch extension to the top, bringing the overall length to 23 feet. ICOM specifies that length as the minimum for 10 to 80 meter coverage using the AH-4. Only about 8 inches of the whip is visible above the copper toilet tank ball that I attached to the top of the mast to make the transition to flagpole complete (see **Figure 2**). A short piece of wire with ring terminals soldered at both ends makes the electrical connection from mast to the finial extension. With the extension painted a nice gold color, the humble origin of the finial is completely disguised.

## The End Result

The flagpole antenna looks very nice in the middle of the front lawn, surrounded as it is by flowering plants installed by Audrey and me. The planting bed is protected from the lawn service's weed whacker by a ring of concrete edging blocks (see **Figure 3**). The plantings also hide what few wires are above ground, completely disguising the fact that it really is an antenna with a flag on it. I purchased a very nice 3 × 5 foot flag from a local flag dealer and 40 feet of 3⁄16 inch rope, along with a cleat to secure the rope about 4½ feet above the ground completing the flagpole disguise. Now to get to the bits that turn this flagpole into an HF antenna.

The AH-4 tuner is mounted to the "back" of the mount, while the tilting part of the mount is on the front. At the point where the mount disappears underground there is a flat steel strap welded across the legs of the structure. This has an electrical ground bus attached to it

Figure 3 — Base of the antenna with the enclosure in place and the surrounding flowers in full bloom.

to accept the wire radials. The ground connection on the AH-4 is also wired to this bus bar. An 8-foot ground rod driven near the legs of the mount is also tied to this point. This ground rod is connected by a large size wire to the ground rod just outside the radio room.

The coax, ground wire, and control cable are buried from the mount to the house about 6 to 8 inches deep. A direct bury rated coax was purchased for this application. The control cable is four conductor cable obtained from a local electronics supply house. I have used this type of cable before to remotely control other auto tuners on ARRL Field Days. A lightning arrestor was inserted in the coax run and connected to the ground rod near the base of the mount rather than at a point just outside the house. The ground rod at the base of the antenna was connected via a #6 AWG copper wire to both the station ground rod and the house safety ground.

## The Radials Finish the Story

The radial arrangement I use cannot be called optimum. The lawn area I used is probably no more than 40 × 30 feet, with the mast just about in the center. The screened porch of our house encroaches on this area somewhat, so the radials were laid and buried where I could. A total of 4 were placed, roughly at 90° intervals around the base, varying from 10 to 25 feet in length. I used about 100 feet of #18 AWG vinyl insulated wire from a local auto parts store. I know that is not nearly broadcast station quality, but with such a constricted site, it was the best I could do. The sod was cut using an electric edger and the wire was placed in the resulting slit trench about 1 to 2 inches below the bottom of the grass. I had originally planned to install about 300 feet of radials, but the work involved turned out to be more than I could tolerate. More may be added later if I suddenly become 25 again. The radials are all connected to the base of the antenna flagpole (see Figure 1) along with the 8 foot ground rod driven at the base of the mount.

The entire base mount is hidden by the simple expedient of taking a surplus kitchen waste basket and inverting it over the base mount, tuner, and the assorted cables. At the top of the waste basket (now the bottom of the enclosure) about 4 inches was cut off to allow the top of the enclosure to sit right on the top of the base mount. A 3 inch diameter hole was cut off center in the bottom of the waste basket (now the top of the enclosure) and several cuts made outward from the hole to allow it to be maneuvered around the base without having to remove the mast first. These cuts were then secured by screwing small aluminum plates over them. If I need to work on the tuner or the connections, the enclosure can be lifted up and rotated so that it sits on the top of the mount while I make any needed adjustments to the system.

Once all this is done, I simply rotate the enclosure back to its normal position and it slides down the mast until it meets the planting bed that surrounds the base. The white plastic of the enclosure matches the white paint of the mast well and, since the flowers around the base have grown a bit, all anyone can really see is the top couple of inches of the enclosure. The enclosure is not waterproof and it was not intended to be. Really, it is just there to cover the tuner and all the wires and cables that attach to it, purely for aesthetic value. If your antenna is in the middle of your front lawn, it pays to think about the aesthetics of the situation.

While I do not spend a lot of time on HF and certainly am not what one could call a contester, I wanted to have an antenna that would let me use the capabilities of my radio to the degree I needed. That statement brings us to the obvious question: "So, how does it work?" Once the labor of installing radials and coax and control lines was done, the testing began.

## So, How's It Play?

I am quite sure that those hams who are lucky enough to have towers and multiband Yagis on big rotators running the legal limit will be quite underwhelmed by the results. But for those of us who, for one reason or another, cannot put up big towers or who have to go into stealth mode due to covenant restrictions, even a half decent antenna is better than none at all.

Given the experience I had with my commercial vertical for the past eight or nine years, I would say the results are satisfactory. With my previous setup and 100 W, I was pretty much able to work most of the stations I could hear and I've found much the same to be true with the flagpole antenna. The addition of coverage on 75 meters is welcome as the previous antenna was only designed to go down to 40 meters. Because I live only a couple of miles from the Gulf of Mexico, the water table is quite high so the performance is somewhat enhanced compared to another location on higher or rockier terrain.

The point of all this is simple. If your site is less than optimum, or if antenna restrictions limit what you can set up, you do not have to forgo your HF privileges. You just have to get a little creative and hide the antenna in plain sight. Just make sure that you pay attention to the details. As someone once said: "Take care of the little details and the big picture will take care of itself."

---

**Notes**

[1]J. Ebner, N8JE, "Flagpole Vertical," *QST*, Apr 2007, p 21.

---

Photos by the author.

ARRL member Geoff Haines, N1GY, was first licensed in 1992 as N1LGI. Geoff upgraded to Amateur Extra in 2005 and received his current call sign. He retired following a career in respiratory care. Geoff currently holds several ARRL appointments in the West Central Florida Section, including Assistant Section Manager, Technical Coordinator and Net Manager among others. He is a past president of the Manatee Amateur Radio Club, and a member of several ham radio clubs both in Florida and Connecticut. In his spare time, Geoff is the editor of the quarterly e-magazine "The Experimenter" for the West Central Florida Section. Geoff is active in designing small projects such as antennas and accessories suitable for the new ham. He also finds time to update his website, **n1gy.com**, on a regular basis. Recently, his wife Audrey became licensed as KJ4YMX. Geoff can be reached at 904 52nd Avenue Blvd W, Bradenton, FL 34207 or at **n1gy@arrl.net**.

# Down Periscope!

## *Keep that vertical out of sight and safe from lightning when you're not using it.*

Burt Yellin, K2STV

I recently came to the conclusion that I required a different antenna in order to take advantage of the upward-bound side of the current sunspot cycle. In order for me to be successful, there were several issues that had to be resolved. First, I wanted an antenna with a low radiation angle. To me, that meant a vertical. Second, I had to abide by my homeowner restrictions and covenants that would not permit buried radials. Third, I live in the Tampa Bay area — the lightning capital of the US. I am very concerned about lightning effects after surviving two lightning strikes at two different locations in this area. The thought of a rod poking more than 20 feet into the sky scares me. My final concern is that I have very little space for an antenna footprint without conflicting with the priorities of my long-suffering wife. My solution — a vertical antenna that is only up in the air when I need it.

### My Vertical Radiator

My band of choice is 17 meters. The antenna I ended up with is a full quarter wave vertical on 17 meters composed of telescoping aircraft aluminum tubes. The total radiator length, after adjusting for best SWR, was about 12 feet, 10 inches. I used three telescoping tubes — 0.875, 0.75, and 0.625 inches in diameter.

### The Counterpoise

The counterpoise is the unique part of this antenna. My counterpoise is composed of two shortened citizens band (CB) antennas. A company named FireStik manufactures a tunable CB antenna that is rated at 650 W [That's a surprising safety margin for a service limited to 12 W PEP. — *Ed.*] and is only 3 feet long.[1] Since I normally run between 500 and 600 W, this product was a natural choice for me. These CB antennas are tuned via a plated ¼-20 screw at the end. While their frequency can be altered by a few megahertz, it was not quite enough to tune to 17 meters. I tried using a piece of ¼-20 rod as an extension, but could not get a sharp enough dip in the 17 meter SWR while electrically lengthening it.

I then tried a "top hat" capacitance end-loading arrangement and got the bandwidth necessary for SSB operation with an SWR of less than 1.5:1. The top hats

The completed vertical in its extended position.

Figure 1 — The capacitance hat end loading arrangement on the repurposed CB whips used as radials.

are 4.75 inch aluminum discs that are held in place by the supplied ¼-20 screws and by ¼-20 stainless steel nuts, washers and lock washers on either side for better electrical connectivity (see **Figure 1**).

## Construction Details

The vertical section is doubly insulated from the mounting plate by shrink tubing over the aluminum. The tubing acts as a space shim to make a better fit in the 1 inch PVC support tube. I then reamed out the stops from the insides of two PVC couplers, slid them on the main piece of PVC, and glued them into place to act as another layer of insulation. They also serve as the U-bolt mounting surface.

The largest tube extends several inches below the PVC pipe. I drilled a 3⁄16 inch hole ½ inch from the bottom of the antenna tube and secured the coax center conductor with a #10-32 stainless steel bolt and lock washers. I used conductive/antioxidation paste on the connection. I then wrapped the connection with tape and put a piece of heat shrink tubing over it to make it weatherproof. The coax braid was secured to the ¼ inch aluminum plate with a #10-32 bolt, lock washers, and conductive/antioxidation paste. **Figure 2** shows the construction details of the vertical.

## Mounting Arrangement

I now had a vertical antenna, but how could I mount it so that it couldn't poke permanently into the sky? I came up with the idea of a crank up pole. I purchased 10 foot and 5 foot lengths of 1½ inch galvanized steel pipe. I then pounded the 5 foot length 3 feet into the ground and dug out a base around this pipe about 1 foot deep. I mixed Sakrete and filled in the space. At the top of the 10 foot length I drilled a 5⁄16-inch hole just below the threads and bolted a 5⁄16 × 3 inch stainless steel bolt to the pipe.

I then mounted the pulley on the bolt and secured it with two stainless steel nuts. After that, I capped the pipe top with a 1¼ inch PVC cap. This entire assembly was then screwed onto the 5 foot length via a 1¼ inch galvanized steel coupler. This was done after I slid a 1½ inch × 12 inch galvanized steel nipple to serve as my *trolley* on to the 10 foot length of the 1¼ inch galvanized steel pipe (see **Figure 3**). This trolley is the means by which the antenna slides up and down the 10 foot pole. I happened to have large diameter shrink tubing and I shrunk it over the cap and the coupler.

## Making it Go Up and Down

The final step was to devise a means for raising and lowering the antenna. This was accomplished through the use of a simple winch. I found a light duty trailer winch with 600 pound pulling capability at an overstock store, but something similar should be available from other sources. Pulling capability was

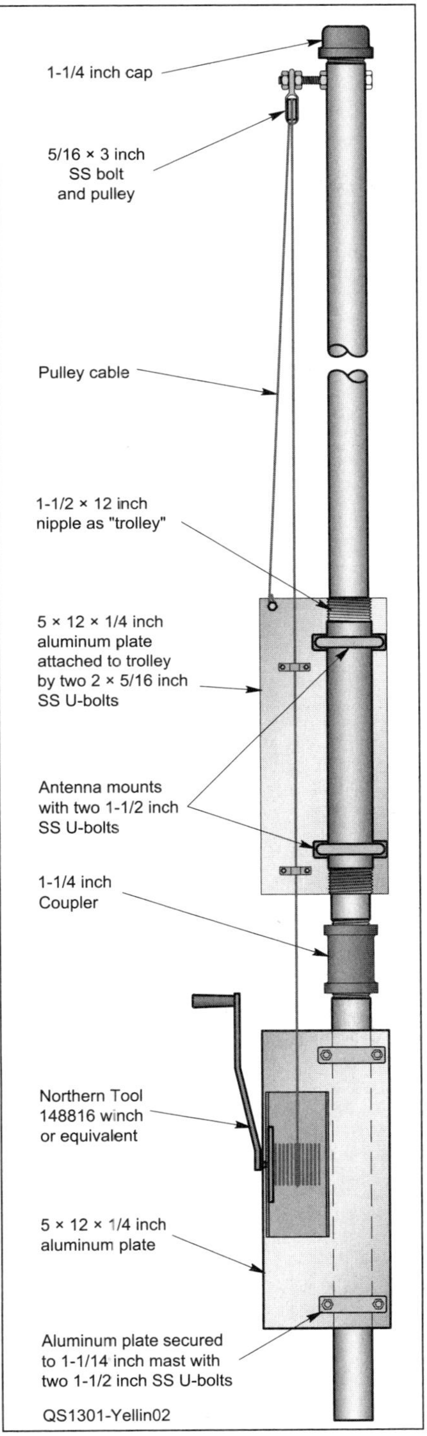


**Figure 2** — Construction details of the antenna and mounting system.

**Figure 3** — View of the trolley on the 10 foot length of the 1¼ inch galvanized steel pipe. This trolley is the means by which the antenna slides up and down the 10 foot pole.

**Figure 4** — The winch used to raise and lower the antenna. It can work in either position, but performs better if raised.

**Figure 5** — The end of the winch cable is attached to the trolley as shown.

**Figure 6** — The fully nested antenna is approximately 13 feet in height.

far in excess of the weight of the antenna, the antenna trolley, the aluminum mounting plate, and all the mounting hardware necessary to move this antenna up and down, thus providing a margin of safety (see **Figures 4** and **5**).

When raised, the antenna is approximately 23 feet in total height. When nested it is approximately 13 feet in height (see **Figure 6**) — hopefully that's low enough to escape a stray lightning bolt. As a side note, galvanized steel pipe comes in 20 foot lengths, thus giving you the additional height if you want it. However, my winch only came with 21 feet of cable and that was my limiting factor in designing and constructing this antenna.

## On the Air

Although this vertical antenna can be used in the nested position, I choose to use it only when it is raised. The ARRL was kind enough to generate and run an *EZNEC* model of the antenna for me and they discovered that it has 1 dB more gain when raised.[2] Of even more importance, upon raising the antenna, the peak elevation angle dropped from 24° to 17°, providing improved low angle coverage.

I am not concerned with direct lightning strikes while the antenna is lowered. The lowered height of 13 feet is well below the roof ridge vent of my house. I also have on my property two well grounded 17 foot pipes that are mounted about 35 feet apart to support a 20 meter dipole. One of those pipes is just visible in the lead photo for this article.

To protect against coupled lightning effects, the coax connects through an ac coupled surge protector as it enters the station. This protector is directly grounded to two 8 foot ground rods near to where the lines enter my shack. The antenna seems to work quite well as I have had many overseas and stateside contacts since it went up.

**Notes**

[1]**firestik.com**, FireStik FS3.

[2]Several versions of *EZNEC* antenna modeling software are available from developer Roy Lewallen, W7EL, at **eznec.com**.

ARRL member and Amateur Extra class licensee Burt Yellin, K2STV, has been a licensed ham since May 1956. It was ham radio that brought him to a career as an electronic engineer. He graduated from Brooklyn (NY) Technical High School in 1956, went to RCA Institutes and, while working as an electronics technician, succeeded in earning a degree at Long Island University by attending at night for 9 years.

You can reach Burt at 1813 Columbine Pl, Sun City Center, FL 33573-5053 or at **k2stv@tampabay.rr.com**.

# The Flagpole Antenna

***This antenna may work around your homeowners association restrictions.***

John Ellis, NP2B

[Jeanette Ellis, NP2C, photo]

Many of us are governed by homeowners association "covenants, conditions, and restrictions" (CCRs) that preclude deploying HF antennas. Rather than argue the point, it's a lot easier to put up an antenna that doesn't offend anyone. This flagpole antenna may be the answer.

## Select a Site

Locate a site away from metallic objects like aluminum downspouts, concrete walls with reinforcing steel, and light fixtures (see the lead photo). Make sure you are not violating any setback rules. If your homeowners association requires approval for its location, by all means get it.

This is a vertical antenna that requires radials, so the more radials, the better. You can hand-trench them, or call a company that installs invisible dog fences. They can quickly put down 1,000 feet of wire, and the next morning you'll never see it.

## Construction

**Figure 1** lists the materials needed. Procure your flagpole kit, the extra 4-foot section, the acetal rod, and the 2-foot length of PVC before you begin construction. Carefully cut the extra flagpole section in half, taking care that you don't bend or dent it. It is used as a spacer, ensuring the acetal rod is rigidly secured within the PVC, yet it can be removed in the event of a storm or for repairs. Place the untapered piece within the PVC, and the acetal rod within this untapered section. Assemble the flagpole pieces first, and make sure everything fits. The inside of the PVC pipe is slightly larger than the outside of the extra flagpole piece that you just cut, and the acetal rod is slightly smaller than the inside of that piece. Everything should fit snugly. This is why it's so important not to deform or dent the flagpole piece that you just cut.

## Working on the Ground

Dig a hole about 30 inches deep and line the bottom with large gravel or small stones to a depth of 6 inches. The hole should be just wide enough to accept the PVC and a couple of bags of ready-mix concrete. Place the PVC, the cut section, and the rod in the hole, and adjust the height of the PVC and cut piece so they are about 1 – 2 inches above ground level. The acetal rod should protrude another 2 feet. Ensure that everything is vertical, and pour the concrete.

Next, install the radial system. A combination of long and short wires is fine, because you're not looking for resonance on any frequency. A good way to tie all these wires together at the base is with a radial plate. A nice one that's resistant to corrosion and made of high-grade stainless steel is available from DX Engineering (**dxengineering.com**). Install a piece of conduit from the shack to the base of the flagpole to accommodate connections to the tuner. I recommend using a conduit rather than directly burying the cables, just in case you have any problems with the connectors or cable.

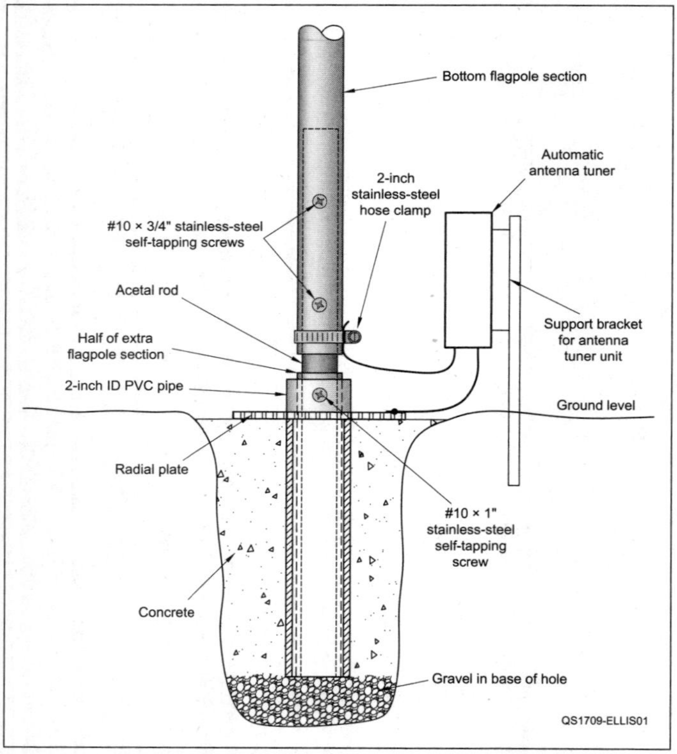


**Figure 1 — Materials needed for the flagpole antenna are pictured, and listed here. [Jeanette Ellis, NP2C, illustration]**

**Flagpole kit with extra 4-foot section — Valley Forge Flag Company model APF20F (valleyforgeflag.com) or equivalent**
**Schedule 40 PVC pipe — 2-foot length, 2-inch inside diameter**
**Acetal rod — 4-foot section of acetal (polyoxymethylene) rod, 1.875 inches diameter, from Interstate Plastics (interstateplastics.com)**
**Antenna tuner unit (ATU) — fully automatic, weatherproof, and capable of tuning a wide impedance range**

## Assembly

Drill a small hole, about ⅛ inch by 1 inch deep, through the PVC that you installed in the ground, through the cut piece, and into the acetal rod. Install a #10 by 1-inch stainless-steel self-tapping screw. This will keep the entire base assembly in its current position. Assemble your five-section flagpole, placing the bottom section over the acetal rod, leaving a ½- to ¾-inch gap between the base and the bottom section of the flagpole. This is your center insulator. Secure the bottom section of the flagpole to the acetal rod, maintaining the gap separation by drilling two small holes through the pole into the rod, and secure with two #10 by ¾-inch self-tapping stainless-steel screws. The #10 screws are strong enough because the pole is very light.

Finally, mount the antenna tuner as close to the base as possible, and install with all electrical connections facing down to avoid accumulation of water, dirt, or debris. Connect the output of the tuner to the bottom section of the flagpole using a stainless-steel hose clamp and a short piece of wire. Don't use the screws you installed to secure the pole to the rod for the electrical connection. Connect the ground of the tuner to the radial plate with a short piece of heavy wire. Pull your cables from the tuner to the ham shack, plug everything in, power it up, and call CQ!

[Remember to pay attention to proper lightning safety grounding of the cables entering the house. See the Lightning Protection sections in *The ARRL Antenna Book* or *The ARRL Handbook.*[1] — *Ed.*]

If you have difficulty tuning on any particular band, try adding the leftover 2-foot section to the top. Also, some folks have had improved 80/75-meter performance by adding a random piece of wire from the top to a nearby structure.

**Figure 2 — The antenna tuner unit is the black box at the base of the flagpole. [Jeanette Ellis, NP2C, photo]**

**NOTES**

[1] *The ARRL Antenna Book,* ARRL Item no. 0444, and *The ARRL Handbook for Radio Communications,* ARRL Item no. 0628, are available from your ARRL dealer, or from the ARRL Store. Telephone toll-free in the US 888-277-5289, or 860-594-0355, fax 860-594-0303; **arrl.org/shop/**; **pubsales@arrl.org**.

John Ellis, NP2B, was originally licensed in 1957 as KNØMMI. He has held the call signs KØMMI, W5RPI, and W4MRJ. He and his wife, Jeanette, NP2C, lived on the island of St. Croix in the US Virgin Islands from 1988 to 2012, where he worked as a federal trustee and served eight terms as ARRL Section Manager from 1996 to 2012. John is currently retired and lives with Jeanette in The Villages, in Florida. You can reach John by e-mail at **john.np2b@gmail.com**.

# A Compact 20-Meter Vertical Dipole for Stealth and Portable Applications

***This antenna earned an Honorable Mention in the 2018 QST Antenna Design Competition.***

Steve Appleyard, G3PND

While thinking about the design of a new antenna for a 30 × 50 foot plot of land, I wanted the antenna to be easily erected/dismantled for both stealth and portable operation. I subsequently decided on a free-standing half-wave 20-meter vertical due to it having a reasonable antenna length and no need for a counterpoise. However, as a 20-meter vertical dipole is about 33 feet high, this would still be outside my design parameters. Therefore, I decided to design a shortened loaded vertical dipole by making the bottom section an almost complete helically wound inductance, and the top section a self-supporting conductor with an appropriate amount of inductive loading.

## Design

As I wanted to keep the total antenna length about 16 feet, I chose the main two elements to be a 10-foot-long, 1¼-inch-diameter PVC tube and a 6-foot-long, ½-inch-diameter copper pipe. Using antenna design software for this project was beyond my capability, so I decided on a trial-and-error process, taking advantage of help from Norfolk Coast Amateur Radio Society (NCARS) volunteers — in particular, Bruce, G4KZT, and Phil, G4PQP.

I began by erecting a full-size tuned 20-meter horizontal wire dipole with the center 7 feet above the ground. I then replaced the left-hand section of the dipole with my experimental bottom vertical section. After several iterations, I arrived at a dipole resonant at 14.128 MHz. Next, I replaced the right-hand dipole section with my first attempt at the top section. I adjusted the coil until I again achieved resonance. This gave me the basic pieces necessary for the vertical dipole.

## Assembling the Compact Vertical Dipole

Originally, I was unable to achieve a low SWR at resonance. However, after much experimentation, I found that I needed to increase the winding separation closer to the feed point and then progressively reduce it down to the final 1-centimeter spacing. The coil winding details are shown in **Figure 2** [For those without a metric ruler, 1 centimeter = 0.4 inches. — *Ed.*]. The main issue is the high voltage at the bottom of the lower helical winding, as maximum voltage occurs at the ends of a half-wave dipole. The risk of arcing was minimized by fitting a 30 A terminal block at the wire end to provide the rounded edges of the terminal barrel.

Using the materials in the "Parts List" sidebar, I began assembly following these steps:

1) Stick a 6 foot × 5 inch strip of masking tape along the length of the 10-foot PVC pipe from the end you designate to be the top of the dipole.

2) Mark the masking tape with each turn of antenna wire as shown in Figure 2, starting from the top of the antenna.

3) Stick double-sided tape alongside the masking tape to keep the antenna wire from slipping.

4) Insert the reinforcing wood into the tube, ensuring a 12-inch clearance from the top.

5) Fix the three-way terminal block with the center contact removed, as shown in Figure 2,

### Parts List

- 10 foot × 1¼ inch PVC pipe
- 9-foot length of wood to insert in the PVC pipe to give it additional rigidity [A 1⁵⁄₁₆-inch-diameter wood dowel, available in 6-, 8-, and 10-foot lengths from a hardware store, perfectly fit the inside diameter of 1¼-inch PVC pipe — *Ed.*]
- 6-inch length of wood with a ½-inch hole drilled vertically through it
- 6-foot length of ½-inch diameter copper pipe
- ½-inch hose clamp
- 15 A terminal blocks, three-way (Bussman TB300-03 or equivalent) and one-way (Cinch 1-142 or equivalent), as shown in Figure 2
- Masking tape, double-sided tape, self-amalgamating tape, and electrical tape
- 50 feet of copper wire; #12 AWG stranded insulated wire was used in the prototype, but smaller diameter wire has been used in subsequent antennas
- Appropriate length of coaxial cable, terminated with a PL-259 connector

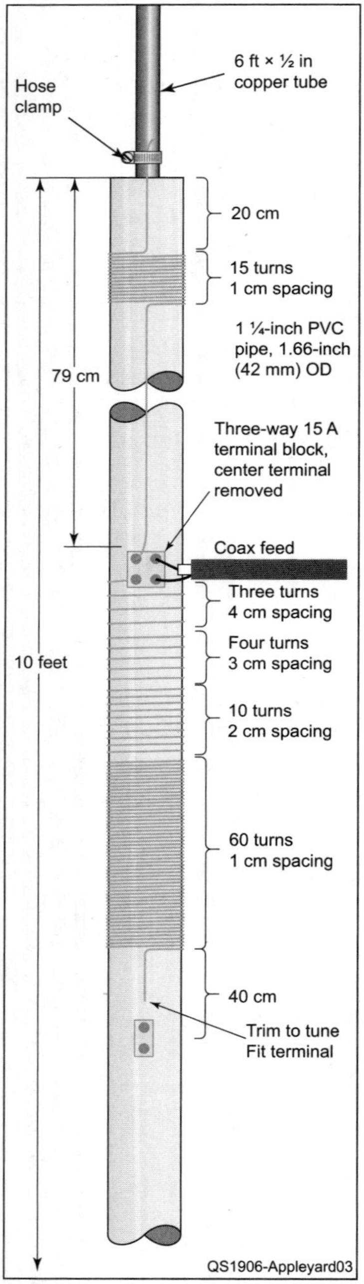


**Figure 2** — **Details of the dual coil assembly.**

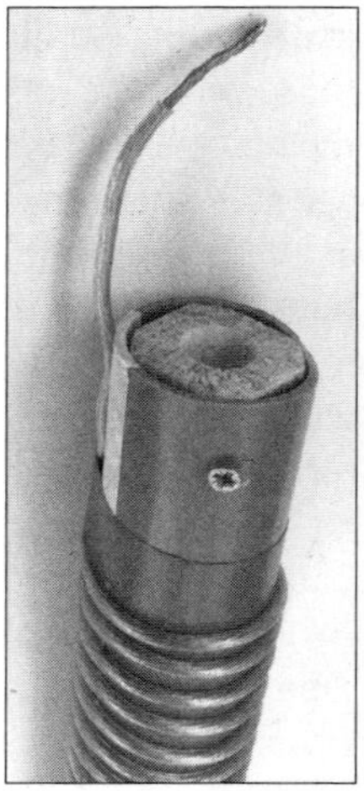

**Figure 3** — **Drilled wooden insert installed in PVC pipe.**

using two wood screws.

6) After taping the wire to the pipe so that 4 inches extends from the end (this end will connect to the copper tube), start winding the antenna wire from the top end of the PVC pipe so that each turn coincides with the markings on the masking tape. Wind it as far as the terminal block and terminate it into the block. Tape the winding every 6 – 8 inches with electrical tape to keep it secure.

7) Terminate the remainder of the antenna wire into the other half of the terminal block and start winding the lower section of the antenna (see Figure 2). Tape the wire at the last turn and then tape a 16-inch straight section along the length of the tube.

8) Fit the 6-inch wooden insert into the top end of the tube and secure with a small wood screw through the plastic tube (see **Figure 3**).

**Figure 4** — **The copper pipe installed.**

9) Insert 6 inches of the copper pipe into the pre-drilled wooden insert and fit the hose clamp with the tail of the previously stripped back wire under it (see **Figure 4**).

10) Attach the coaxial cable to the terminal block with the shield connected to the bottom section of the antenna. Make one turn of the coaxial cable around the tube at the center of the terminal block and tape to hold.

## Tuning the Antenna

Mount the antenna, using an umbrella base, at least 10 feet from nearby objects. The coaxial feeder should be brought away perpendicular to the antenna for at least 3 feet before it can drop vertically. As the feed is

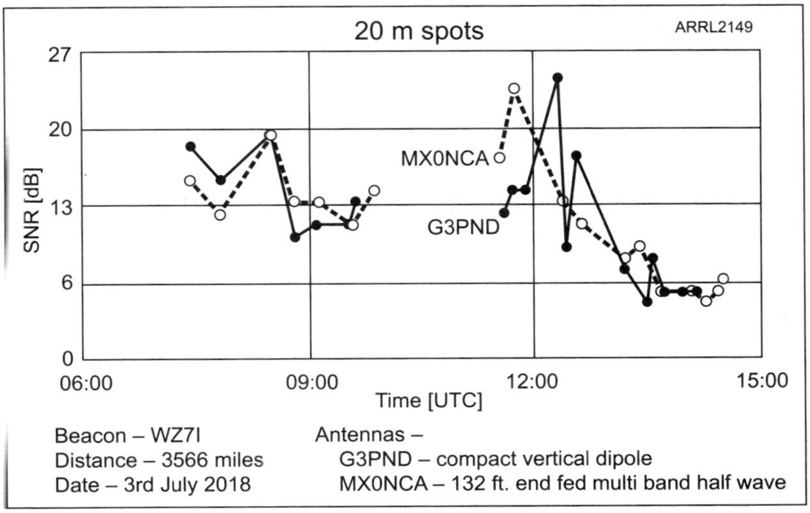


**Figure 5 — An antenna comparison conducted on July 3, 2018.**

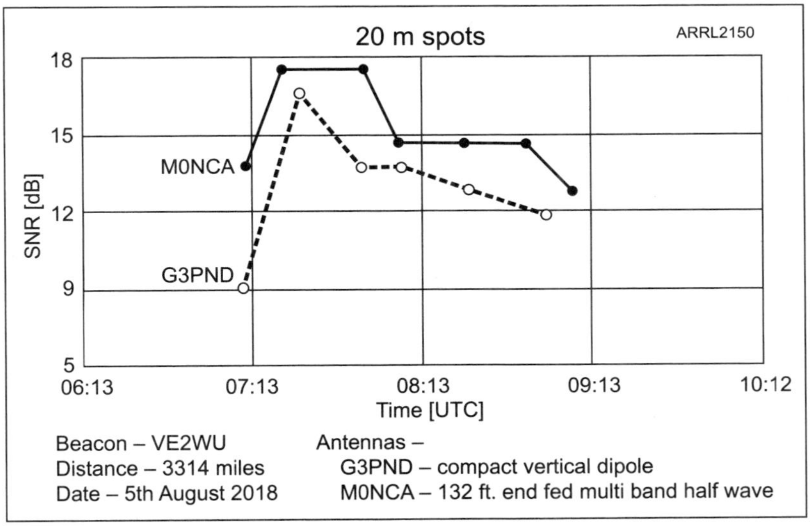


**Figure 6 — An antenna comparison on August 5, 2018.**

unbalanced, a 1:1 choke balun was used to suppress any common-mode current during testing. Subsequently, I have not bothered with the balun as the common-mode current doesn't appear to cause any problems, at least up to 100 W.

The antenna is tuned by trimming the bottom straight-wire section. The length is longer than required, so the antenna will be resonant below 14 MHz. It should not be necessary to trim more than 8 inches off this tail — it is important that at least 8 inches remain. If resonance is still below 14 MHz when 8 inches has been trimmed off, remove a turn from the bottom of the coil. At resonance, the SWR should be below 1.2. If it is not, then a small adjustment can be made to the top section by releasing the hose clamp and dropping the copper tube by a small amount. Strip back ½ inch of insulation from the bottom of the antenna and fit a single-way 15 A terminal block.

Once tuning is complete, replace or cover the temporary electrical tape with self-amalgamating tape. This should also be put liberally around the terminal blocks and the hose clamp. The open top of the copper tube should also be made waterproof.

The final result is an antenna with an SWR of less than 1.2:1 at 14.128 MHz, and less than 2:1 across the full 20-meter band. Several antennas have subsequently been constructed with almost identical results. I have tested the antenna up to a maximum of 400 W, measuring the temperature at various points. There was no noticeable temperature rise.

**Warning** — High voltages exist at the two ends of the antenna. If there is any risk of a person touching the copper tube, a non-conductive plastic pipe can be placed over it. A larger diameter, non-conductive tube can also be fitted over the bottom section of the dipole (see the lead photo). These tubes will only have a minimal effect on the SWR of the antenna.

## On-Air Testing

I am a great advocate of using the Reverse Beacon Network (RBN) to compare antennas.[1] I currently have a 20-meter horizontal dipole with its center at 20 feet and a 130-foot multi-band end-fed antenna. Both antennas perform well given the conditions at the time. Tests on the compact vertical dipole were carried out from June 30 – August 5, 2018. In the tests, I used my call G3PND for the compact vertical and the NCARS calls MXØNCA and MØNCA for my existing antennas.

Given its small size, the compact vertical-dipole performed amazingly well. For the most part, it matched the performance of the dipole and end-fed antennas and, on occasion, it was the best of the three. **Figures 5** and **6** show a couple of the days during the testing period using the RBN Spot Analysis tool. The results from all the beacons that responded to the test transmissions are retained by the Reverse Beacon Network and can be accessed at **reversebeacon.net**. To do this, click **MAIN** on the opening page, then **dx spots**, and fill in the date, the region, and the call signs to compare (e.g. G3PND, MXØNCA).

This antenna lends itself to further experimentation, such as determination of the optimum height above ground and versions for other HF bands.

**Notes**

[1]S. Appleyard, G3PND, "Using the Reverse Beacon Network to Test Antennas," RadCom, Jun. 2018, pp. 26 – 30.

Photos by the author.

Steve Appleyard, G3PND, was first licensed in 1961 while studying electronic engineering in college. He went on to work on the development of communication and navigation systems with the Marconi International Marine Company. In later years, he became the CEO of a number of electrical engineering companies. Now retired, he is secretary of the Norfolk Coast Amateur Radio Society. Steve edited the book International Antennas, published in 2017 jointly by ARRL and the Radio Society of Great Britain (RSGB), and he wrote the first chapter, "Using the Reverse Beacon Network to Test Antennas." He is co-author of Getting Started in EME, also published by the RSGB. You can contact Steve at **sfappleyard@btinternet.com.**

# Double-Topped Flagpole HF Antenna

*This multiband design received an Honorable Mention in the 2018 QST Antenna Design Competition.*

Don Crosby, W1EJM

The double-topped flagpole HF antenna (DTFA) covers 80/75 meters to 10 meters (except 30 meters), often with the use of only a transceiver internal tuner. This antenna uses a vertical mast/flagpole as the feed line and two wires of unequal lengths to resonate the antenna on 75/80 and 40 meters (see **Figures 1** and **2**). The DTFA functions well with wire angles between 60° and 180° and over a variety of radial footprints. A key benefit of the design is that it avoids using highly visible coax or window-line feeder, insulators, baluns, isolators, or traps.

## Designing the Antenna

*EZNEC* software[1] allowed me to determine the flattop lengths for my first prototype antenna. Its performance is shown in **Figure 3**. During further testing, there was good agreement between the computer models and the experimental data measured on four installed antennas. Various lengths for the top wires were explored using NEC simulations. Two unequal length wires could provide a six-band match on 75, 40, 20, 17, 12, and 10 meters.

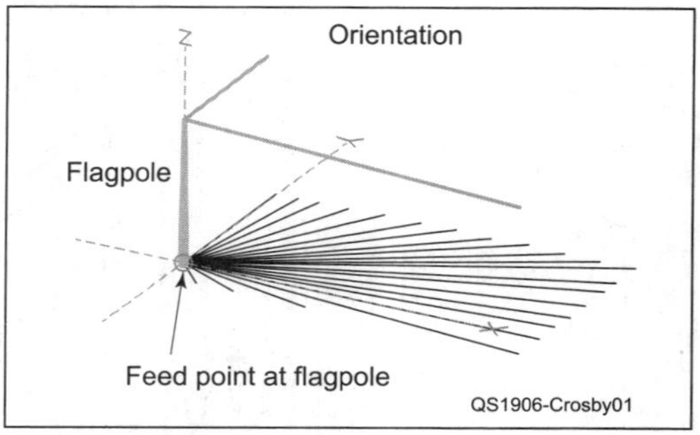


**Figure 1** — The flagpole location. The thin horizontal wires are nearly invisible.

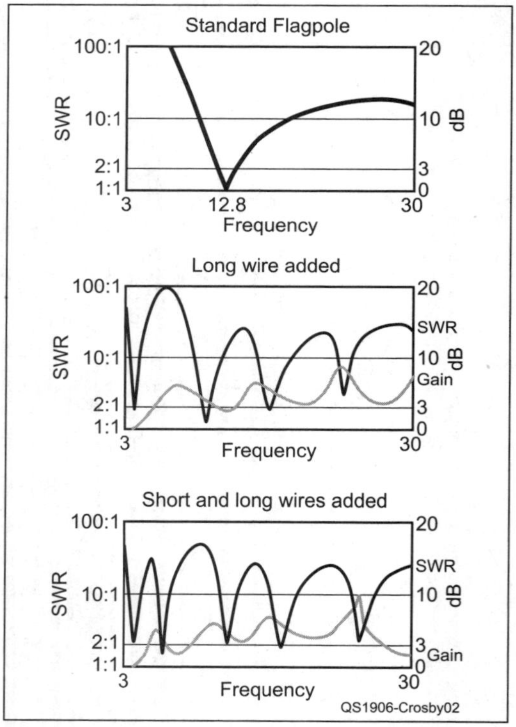


**Figure 2** — The left curve shows a 20-foot flagpole with one SWR dip at quarter-wave resonance for a gain of 0 dBi (or less) on all bands. The middle graph shows the attachment of a 50-foot wire, creating four SWR dips (black curve), one in the 75/80-meter band and others harmonically related, with a gain (gray curve) rising above 2 dBi on all bands above 5 MHz. On the right, 25-foot and 50-foot wires are attached, and a fifth SWR dip appears near 7 MHz, gain rising above 2 dBi on all bands above 5 MHz.

There are four installations of the DTFA near my locale, each having variations in the mast/flagpole height and the radial footprints. Those variations are accommodated by trimming the flattop wires. See the "*QST* In Depth" web page, **arrl.org/qst-in-depth**, for the extensive test details and radial footprints.

For the 2018 *QST* Antenna Design Competition, maximum dimensions specified a 30 × 50 foot plot size with a 30-foot height limitation. The engineering test version of the DTFA used mast heights of 20 feet and 25 feet. The horizontal wires are not noticeable, being 0.03 inch in diameter. They are nominally 49.7 feet and 22.5 feet and ranged in separation angles of 30° to 180°. These thin wires handle the SSB legal-limit power with only about 0.2 dB in wire loss.

The engineering test was conducted with 24 radials on ground within the competition-mandated 30 × 50 foot area. The mast/flagpole is insulated and fed at the base. Three driven ground rods 8 feet in length are also at the base.

## Testing the Prototype

I standardized a 20-foot mast/flagpole height for use in the installations (see **Table 1**) where the antenna wires had an 85° to 110° angle between them to conform to the available plot size. I made my mast/flagpole from 2-inch-diameter aluminum with a .06-inch-thick wall that acts as a feed line for the horizontal wires. However, the mast can be any 20- to 25-foot-long conductive pole that is insulated from ground at the feed point, or a non-conductive mast with #16 AWG insulated wire as the vertical electrical member. The insulated base mounting is similar to that used in a previous project.[2] Stainless-

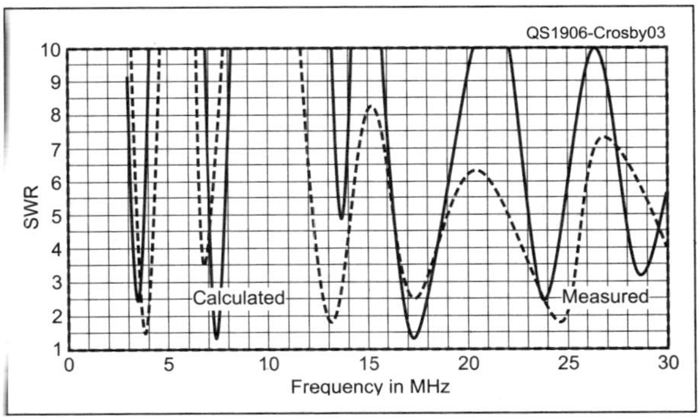


**Figure 3** — The SWR at the base of the flagpole, as calculated by NEC, is shown by the solid line. Actual measured values are shown by the dotted line.

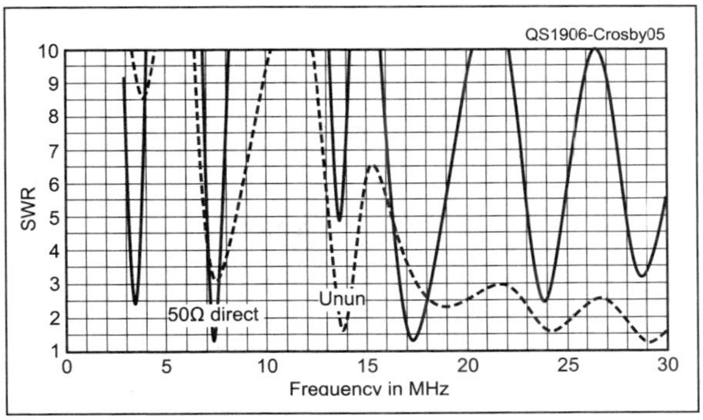


**Figure 5** — The SWR feeding directly from a 50 Ω feed line is shown in the solid line. The SWR as seen through a 50 to 200 Ω unun is shown in the dotted line.

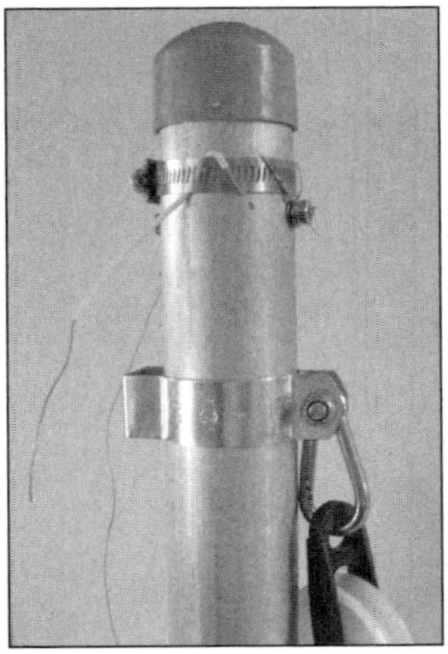

**Figure 4** — The thin wires attach to the pole with a hose clamp, using soft sleeves to protect the wires.

steel hose clamps are used for connecting the horizontal wires to the mast (see **Figure 4**). Strain relief of the wires is important for reliability. Ideally, the wires are routed through 2-millimeter ID Teflon™ tubing, entering on the low side of the clamp. The bare wire ends must contact the mast. In use, the wire exits the clamp in the direction the wire would tend to run.

About 500 total feet of radials were installed within the available space. In three cases, the radials were installed in existing lawns by using an electric string trimmer to "scratch" a notch to the depth of the grass-earth level and the insulated radial wires were installed in the notch and held in place with radial anchor pins.

In the first prototype, I used 60 feet of direct burial 50 Ω coax with the center conductor connected to the mast/flagpole, and the sheath to the radials. The internal tuners in my transceivers (an Icom IC-7300 and Kenwood TS-570) were able to tune most bands (see **Figure 3**), but I opted to add a 1:4 unun as a step-up transformer (relay selected) to improve the match on the high bands where the antenna impedance is nearer to 200 Ω, as shown in **Figure 5**. The unun allows higher power operation by lowering the SWR and the losses on the coax. A remotely located auto tuner is another good solution but is costly if using legal-limit power.

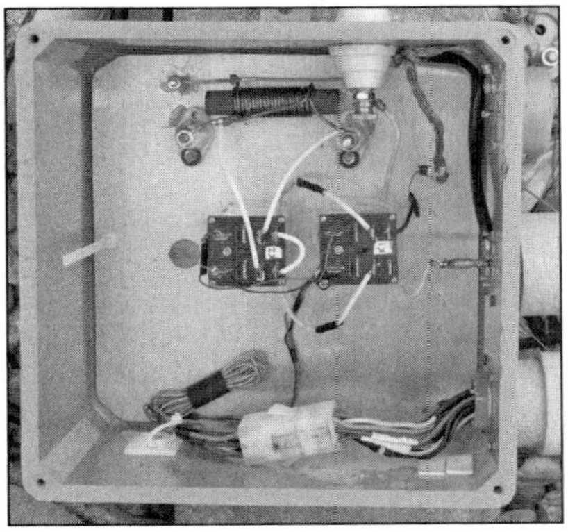

**Figure 6** — The relays and step-up transformer are housed in a weatherproof box.

I used double-pole, double-throw power relays with 12 V dc coils and housed them with the step-up transformer in an 8 × 8 × 4 inch weatherproof box, available at most hardware stores (see **Figure 6**). The schematic of the remote switch box and control switch is shown in **Figure 7**.

I used a RigExpert AA-55 Zoom Analyzer to collect data for various heights, angles, and wire lengths. The measured data confirmed that the angle between the wires can be varied between 180 degrees to as little as 60 degrees with minimal effect on the SWR curve. The ends of the flattop wires could be bent down to make the antenna fit into a smaller area. However, to prevent persons or pets from contacting a high voltage, wire ends would be kept at least 8.5 feet above the ground level (see the sidebar, "Safety Caution"). DTFAs are installed in our community as listed in Table 1. There are slight differences between them due to property limitations, but the performance is similar.

## Safety Caution

Because high-voltage points and radiating parts of this antenna are close to areas where persons or pets might be present, danger from RF radiation must be considered. Instructions for doing this evaluation can be found at **arrl.org/rf-exposure**. In addition, because direct contact with RF voltages can cause painful burns, unless provisions are made to prevent persons or animals from approaching the antenna during operation, the flagpole should be insulated up to a height of 10 feet and the ends of the wires should not be reachable from the ground. Place a low wall — such as that shown in the lead photo — to prevent close approach to the flagpole. This wall will conceal the switch box and insulating mount from view and make contact more difficult.

**Table 1**

*Installed Dimensions, Feet*

| *Location* | *Pipe height* | *Angle* | *Long wire length* | *Long wire end height* | *Short wire length* | *Short wire end height* | *Wire orientation* |
|---|---|---|---|---|---|---|---|
| Site 1 | 21.25 | 110° | 45.1 | 8.5 | 24.2 | 8.5 | Slanted down |
| Site 2 | 20 | 80° | 46 | 17 | 27 | 17 | Slanted down |
| Site 3 | 25 | 180° | 49.7 | 19.7 | 22.5 | 15 | Drooping |
| Site 4 | 20 | 90° | 50 | 10 | 26.5 | 20 | Long droops, short flat |

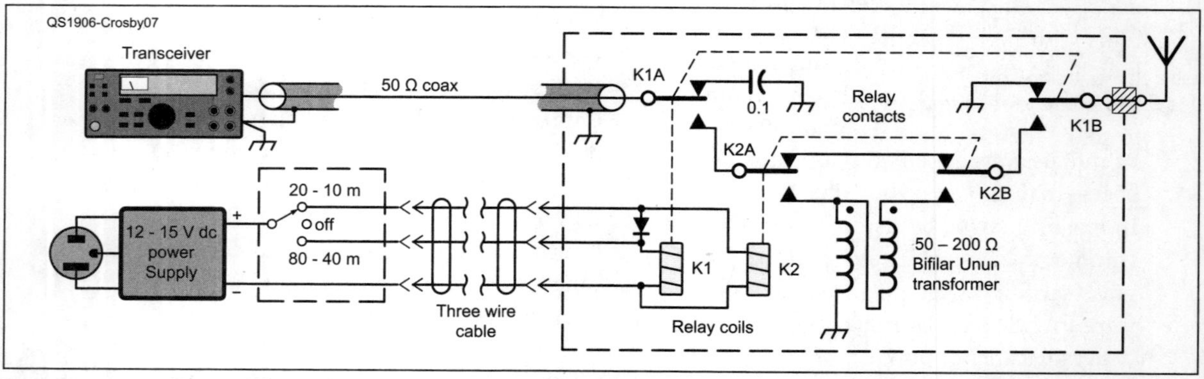


**Figure 7** — Schematic diagram of the remote switch box and its control switch in the shack. The antenna and transceiver are grounded when the switch is off. The 80- to 40-meter position connects directly to the antenna. The 20- to 10-meter position connects via the 50 to 200 Ω transformer. Relay contacts are shown in the off position.

## Trimming Procedure

1. Start with the 60-foot and 30-foot wire lengths. Lowering the wire at the far end, adjust the length and raise the wire back to position.
2. For the 75- or 80-meter frequency tuning, use an SWR meter to find the dip frequency, and write down the frequency and the wire length. Trim the long wire by folding it back on itself about 2 feet. Then find the new dip frequency and write that down. Continue until you are close to the desired dip frequency. Note that you should save your data in case a wire breaks someday.
3. For 40-meter frequency tuning, use an SWR meter to find the dip frequency, and write down the frequency and the length. Trim the short wire by folding it back on itself about 1 foot. Use the SWR meter to find the dip frequency and write it down. Then fold back 1 more foot of the short wire, and find and record the new dip. Continue until you are close to the desired dip frequency. Save the data.
4. Check the 75-meter frequency. It should not have changed by more than 10 or 20 kHz. If it is too high, unfold about 1 foot from the long wire and measure again. Repeat until you are on frequency on 75 meters. Save the data.

## Calculated and Measured Performance

I compared the two-wire NEC model to the actual measured SWR for the prototype antenna using measurements from the Rig Expert AA-55 with a 50 Ω reference. The good agreement of the curves shown in Figure 3 showed that the antenna is working as expected. On 75/80 meters, the antenna functions with high currents in the mast and the long horizontal wire as a low current flows into the short wire. The longer wire length is used to set the center of the desired 200 kHz bandwidth within the 75/80-meter band. As with most 75/80-meter antennas, the band is wide compared to antenna bandwidth of about 200 kHz for low SWR. My in-shack tuner (Palstar AT2K) allows tuning anywhere in the 500 kHz band limits. On 40 meters, high current flows in the vertical mast and on to the short wire, which is used to resonate the antenna on 40 meters while less current flows on the long wire. The antenna is broadband on 40 meters, so I trimmed the short wire to the middle of the band (see the sidebar, "Trimming Procedure").

For both 75 and 40 meters, a number of patterns were run on NEC with angles between the two top wires varying between 60 and 180°. The 80/75-meter DTFA patterns were nearly omnidirectional with radiation at both high angles for NVIS and lower angles for further distances. The azimuth and elevation angles did not vary significantly between the various patterns. For intermediate distances, the 75-meter signal will be about 0.5 dB stronger in the direction of the short wire. On 40 meters, the DTFA showed a broad pattern steered by the longer wire with about a 5 dB front to side ratio.

We conclude that the design is very tolerant of various wire angles and they can be chosen to fit on the property and to use available supports at the far end. The antenna is also tolerant of some variations such as lower end heights, droop in the wires, or a shorter metallic pole.

Many patterns for 20 – 10 meters were plotted from NEC with a 90° angle between the two top wires. The antenna exhibits multilobe gain and notch patterns on the higher bands. These patterns are consistent with off-center-fed half-wave or longer dipoles with more energy at lower takeoff angles. The gain increased from near 0 dBi to as much as 7 dBi, increasing with frequency. The highest gain lobes were in somewhat random directions. Plots of the patterns produced by these tests can be found on "*QST* in Depth" web page at **arrl.org/qst-in-depth.**

### Notes

[1] I used *4nec2* and *Nec2Go* antenna modeling programs for detailed modeling and used EZNEC for display of currents in the wires.

[2] A similar antenna using this same flagpole is described in the May 2018 issue of *QST*, pp. 46 – 49. Details of the insulated base mounting are pictured there.

Photos by the author.

Don Crosby, W1EJM, first licensed in 1956 as KN2VVN, changed call district, and was assigned W1EJM in 1967. Don, a member of ARRL for 53 years, earned a BSEE from Fairfield University (BEI) in 1973 and holds two patents. Don's career was in the satellite optical system area, working on programs such as the pre-Apollo Lunar Orbiter mapping satellite cameras, the Hexagon Big Bird KH-9 Reconnaissance satellites, and the Hubble Space Telescope. Now retired, his ham interest has focused on stealthy HF antennas. Don and Marilyn reside in The Villages, Florida, have been married 56 years, and have two adult children. You can contact Don at **w1ejm@arrl.net** or through his website, **w1ejm.com**.

Don would like to thank Ken Olmstead, KA1NNR, for his contributions to this project. Licensed in 1985, Ken held a career in electronics and computer support, and is a hands-on problem solver. He provided hours of support on this and other antenna projects as a technical sounding board and measurement assistant.

# Compact Stealth Five Band Inverted L Antenna

*If you need to keep a low antenna profile, this may be a solution for you.*

Lorraine Wilson, W1AR

Santa Clara Valley (SCV) is in the San Jose-San Francisco Bay area, popularly known as Silicon Valley in Northern California. This nickname comes from the region's large number of silicon chip innovators and manufacturers, and the many high tech businesses in the area. Technological innovations began here with innovative development of such devices

Figure 1 — View of the installed antenna system in the lower patio (highlighted). Note the trunk of a 60 foot palm tree in the foreground that could not be used for a dipole due to covenant restrictions. The large trees in the far background could not be used either.

**Figure 2 — The horizontally mounted portion of the antenna, as seen from inside the patio.**

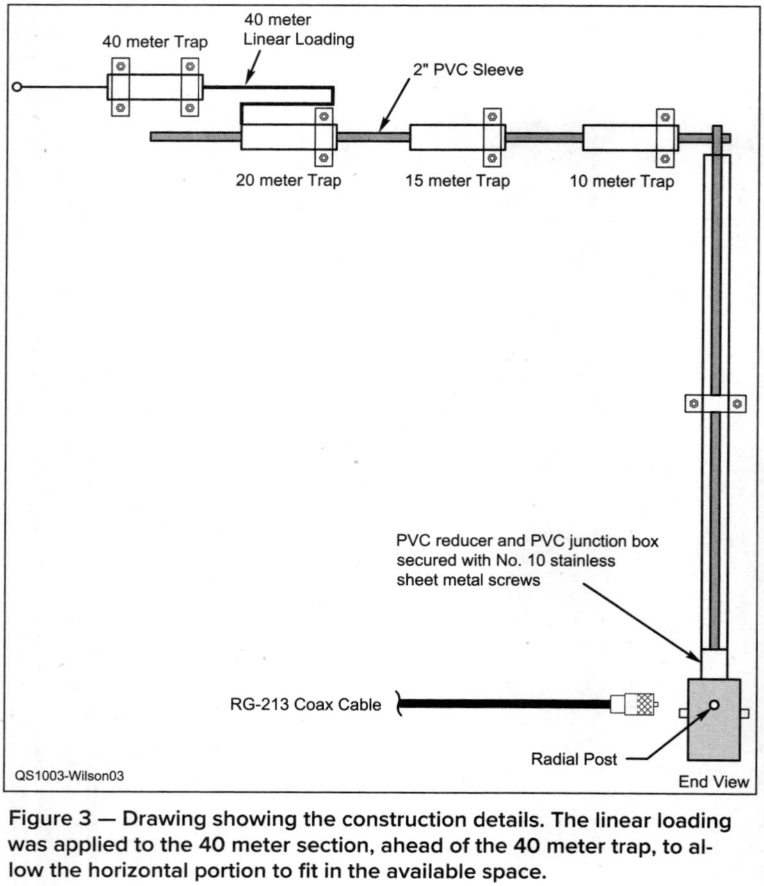


**Figure 3 — Drawing showing the construction details. The linear loading was applied to the 40 meter section, ahead of the 40 meter trap, to allow the horizontal portion to fit in the available space.**

as the radio and the amplifier using new materials made of silicon, including transistors, the microchip, and the home computer.

I recently moved here from Connecticut and found myself in a modest ground floor apartment with an 8 × 12 foot patio and a garden with 60 foot palm trees just outside the front door. There is a covenant restriction (CR) written into the lease forbidding outdoor dish antennas. Wonderful, I thought — a tempting invitation to install a G5RV dipole or an inverted V antenna high up in the palms connected to my HF radio station set up in the living room. Outside the back bedroom were very large trees of all

**Figure 4 — View of the antenna base. Note the stud that will connect to the ground system.**

Figure 5 — The screen mesh is triple-wrapped in weed barrier plastic sheets, and covered with a 1 inch layer of top soil for concealment. The mesh is secured in the concrete patio area by double layers of aluminum aircraft aviation tape.

Figure 6 — Finished surface over ground system with top soil covering plastic sheets.

descriptions, and I thought I had found a place that offered a good antenna location for 80 to 10 meter HF operations.

## The Ugly Truth

I was wrong. After moving in, I was told that this CR also applied to any outside antennas, because they might possibly be visible to neighbors. Drat — time to come up with compact stealth mode antenna options to allow having an antenna installation outdoors. There was another apartment above me, so my choices were limited to something that would fit within an 8 foot high ceiling directly above my patio.

## A Plan is Hatched

I decided to purchase a Hustler five band trap vertical (5-BTV) and make some modifications so it would function as an inverted L trap antenna within the limited area of my patio.[1] The idea was that I would use as much of the antenna vertically as could fit under my patio ceiling, and then transition to horizontal for the balance of the antenna. Other brands could be used if they are capable of being bent at a right angle at around the 10 meter trap point.

## Design Approach

At first, I set up the antenna in the bedroom with just a few counterpoise radial wires spread out across the carpeting, and it worked. With just 100 W of output power, and no sunspots, I effectively managed to work a total of 36 states in the continental US, as well as Mexico, Canada, and Hawaii during several contest trial runs using this suboptimal antenna arrangement.

## Ground System

An effective ground system is important for efficient operation of an electrical quarter wave antenna such as the 5-BTV. Higher efficiency will result in both higher effective radiated power and improved signal strength on receive. [See the article by Rudy Severns, N6LF, on ground systems in this issue — *Ed.*] Quite literally, the ground or counterpoise acts as the missing half, the other quarter wave, of a quarter wave monopole antenna system. [While a copper plated rod pounded into the soil can work, other kinds of ground arrangements can result in more than twice the signal strength, as shown in Rudy's article. — *Ed.*]

Installing the antenna outdoors with a ground system of many buried wire radials would not be feasible in my garden. After modeling the antenna, I concluded that I could get reasonable results using a length of 24 inch wide galvanized chicken wire as an insulated floating (electrically isolated) counterpoise augmented by a solid earth ground rod and a few buried linear loaded wires as radials. This has worked out well for 80 through 10 meter operation.

## Inverted L Radiator

The need to bend the antenna portion into a right angle at the 8 foot point is self-evident, since that is the maximum height of the patio ceiling. Ideally, it would be best to have it oriented to run straight up in the clear, perpendicular to the ground surface. Unfortunately that is not possible in my installation. Something is certainly better than nothing, and I'm pleased by the surprisingly good results.

---

**Notes**

[1]new-tronics.com/main/html/base_hf_5_band.html.

---

ARRL member and Amateur Extra class licensee Lorraine Wilson, W1AR, was first licensed in 1958 and primarily operates CW. She joined the Military Affiliate Radio System (MARS) and used available surplus radio gear at her station, obtaining practical electronics training in order to provide the required 28 V dc systems.

Lorraine's interest in electronics led her to college studies in engineering and business and she now serves as a technology consultant, recently moving her operations from shoreline Connecticut to the San Francisco Bay area. You can reach Lorraine at PO Box 111656, Campbell, CA 95011 or at **w1ar@arrl.net**.

# A Stealth Antenna for HF

## *This 160 meter — 10 meter antenna blends into the neighborhood and won't break your budget.*

David A. Cornell, W9LD

I moved into a new housing subdivision where utilities were underground and there were no visible wires or elevated structures of any kind, except for chimneys and a few flagpoles. The site had been an old rock quarry, so few tall trees were available for supporting antennas. Neighbors were friendly, but they kept lovely gardens and valued their view of the Olympic Mountains. You may wonder what sort of unobtrusive antenna you could install for HF operation in such a location.

My antenna solution for running about 100 W on 160 through 10 meters minimizes visual impact on the environment and fits my criteria of: multi-band operation, reasonable efficiency, robustness, and stealth. I experimented with dipoles, Zepps, inverted Vs, and verticals, then settled on an inverted L system, which uses a gable-mounted 20 foot aluminum flagpole as part of the vertical radiator of the antenna. **Figure 1** shows the antenna on the house from an adjacent street. It looks just like a gable-mounted flagpole.

**Figure 1 — Visual impact of the antenna is that of a flagpole on the gable.**

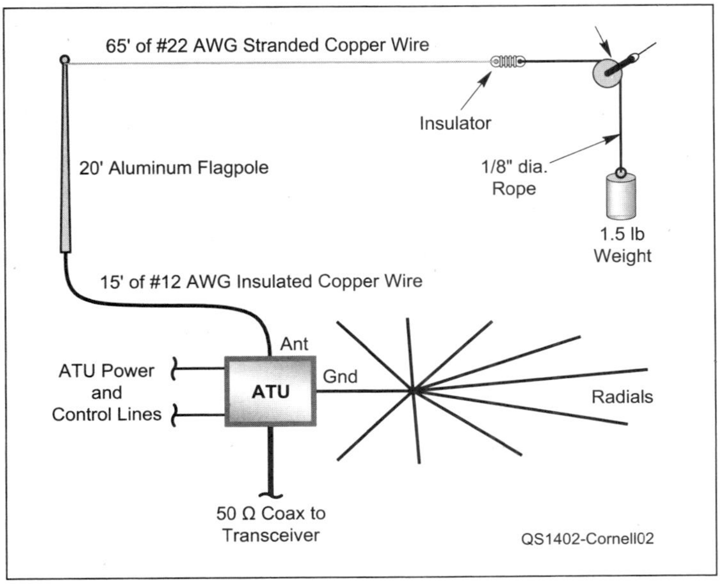


**Figure 2 — Inverted L system includes vertical and horizontal radiating wires, ground radials, and an ATU.**

**Figure 3** — Use a manual edger tool (left) and a push bar (right) to bury radials.

## Antenna Components

The diagram in **Figure 2** shows the ground radials, antenna tuning unit (ATU), and the two upper radiating legs of the L antenna. The 100 foot total length of the upper radiator helps facilitate operation up to Top Band (160 meters). Because of the topography of the lot on which my house is located, I placed the system close to the back of the house. That limited how I could fan out ground radials, but enabled the use of a gable mounted flagpole as part of the antenna.

Just a few inches of sod and surface dirt cover the underlying (poorly conducting) quarry rock, which I improved with six radials of lengths varying between 70 to 110 feet. I installed a total of 500 feet of radials that fan out in six directions from the node at the back of the house.[1] They wind around the base of the house along both

### RF Exposure Safety

Amateurs are — without exception — required by the FCC to ensure compliance with RF exposure regulations. This antenna system is unusual in configuration and installation, thus performing an RF exposure assessment is prudent. Every wire above ground contributes to RF radiation, including the top L wire, the flagpole, the wire under the gable and on the porch, and the wire below the ATU to the node point of the radials. Because the antenna system is on premises entirely within the control of the amateur operator, most likely compliance to the "Controlled" distances is adequate. You can estimate a "worst case" compliance distance, measured from any portion of any wire, by applying the generic General-Purpose Table in RF Exposure and You by Ed Hare, W1RFI.[A] Alternatively the FCC permits Amateurs to use calculations from valid field-strength formulas and principles, or computer modeling using programs based on accepted algorithms such as NEC (Numerical Electromagnetic Code) or MININEC. These more accurate techniques can result in a more precise compliance distances that often are closer to the antenna. Kai Siwiak, KE4PT, in discussing a similar inverted L antenna contrasted several RF assessment methods in his *QST* article.[B] — Ed.

[A]Available from your ARRL dealer, or from the ARRL Store, ARRL order no. 8217. Telephone toll-free in the US 888-277-5289, or 860-594-0355, fax 860-594-0303; **arrl.org/shop/**; **pubsales@arrl.org**.

[B]K. Siwiak, KE4PT, "All Band Attic Antenna," *QST*, Oct 2007, pp 33-37.

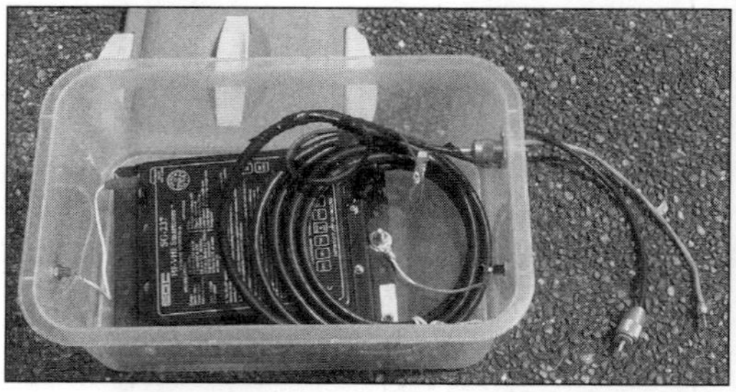

**Figure 4** — The ATU fits inside a protective box for weather protection. Connectors for antenna, ground, coax from radio, and ATU connector mount on the box ends.

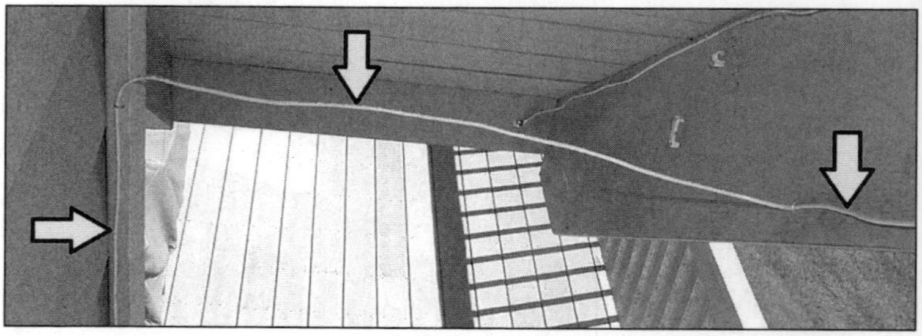

**Figure 5** — Lower part of vertical antenna radiator runs along the house molding and inside the gable as shown by yellow arrows.

sides. **Figure 3** shows Larry Hanson's, K7EKD, method of installing radials just beneath the turf by using a manual edger and a push-tool.

I located the antenna feed point below the deck at the rear of the house. The 100 foot radiator is a non-resonant length [Avoid lengths that are near any multiple of a half wavelength. — Ed.], so I chose an automatic antenna tuner unit (ATU) to match the 50 W cable from the transceiver to the antenna. My ATU automatically provides a match on 160 meters through 10 meters. It resides in a covered plastic box for weather protection.[2] I cut holes in the box for an SO-239 coax connector, ATU power (RCA jack), and antenna and ground (red and black banana jacks shown in **Figure 4**).

## Construction

The vertical L portion of my antenna is 35 feet long, of which the upper 20 feet is a 2 inch OD aluminum flagpole.[3] The remaining 15 feet of #14 AWG solid insulated wire runs along the wooden siding and inside the eave, as pointed out by arrows in **Figure 5**, from the flagpole to the below-deck ATU. That wire need not be straight nor perfectly vertical. I fastened the wire to the house and deck taking care to preserve stealth for deck users [See the sidebar, "RF Exposure Safety." — *Ed.*]. Homebrew brackets support and affix the flagpole to the wooden gable top. **Figure 6** shows how the flagpole is insulated from the bracket by plastic pipe telescoped inside the pole. Secure the junctions between pole sections by drilling and installing three stainless steel sheet metal screws at each junction. Scrape the anodizing coating on the flagpole with a knife for a good electrical connection.

I wired a gas-filled flashtube, for lightning protection, to the base of the flagpole (seen in Figure 6). The ground terminal of the flashtube is wired to a #6 AWG cable connected to earth ground. To drain static charge from the antenna, I paralleled a 1 MΩ 1 W resistor across the lightning flashtube.

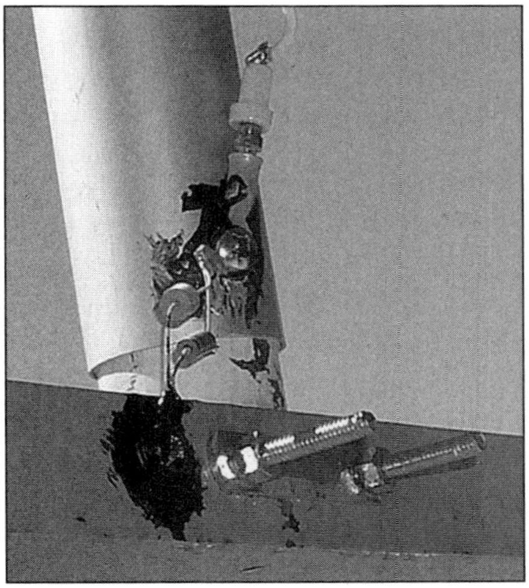

**Figure 6** — The aluminum flagpole mounts to the house gable with homebrew brackets made a steel angle and U-bolts. Insulating plastic pipe telescopes inside the base of the flagpole and stiffens the pole at U-bolt locations.

A horizontal wire, 65 feet of #22 AWG stranded insulated copper, connects to the top of the flagpole to extend usability of the antenna to 160 and 80 meters. In Figure 1 that slender wire is stealthy and all but invisible, extending down and to the left from the top of the flagpole. The tension in the wire is only 1.5 pounds, which minimizes flexure of the flagpole. I soldered the wire to a lug that I secured to the flagpole top using stainless steel hardware. The far end of the horizontal wire (see Figure 2) is tied to a ⅛ inch diameter rope that runs over a tree-mounted pulley to a 1.5 pound counterweight.[4] This weight keeps the tension constant, and allows the wire to sag about 12 inches. This protects the wire from breakage while the tree moves in the wind.

## How it Performs

While one cannot expect this antenna to rival a directional beam, it does enable easy contacts with 100 W on the HF bands. In a single 24-hour period I initiated contacts on 160, 80, 40, 30, 20, and 15 meters with a single call. I've received S9 signal reports from adjacent states on 160 meters. One station commented that I had "one of the louder signals" on a 75 meter net.

The flagpole portion maintains a low profile and is the only visible evidence of my inverted L antenna. My antenna has already withstood the seasonal cycle of Seattle-area winds. If you want to operate the HF bands and prefer to minimize visual impact in the neighborhood, consider this inverted L antenna system so you can enjoy Top Band through 10 meter band operation.

**Notes**

[1]Stranded insulated #14 AWG copper wire available from various home improvement stores.

[2]Suitable boxes available from art craft stores.

[3]20 foot aluminum flagpoles available from **United-States-Flag.com**.

[4]Boating hobbyists may have surplus "zincs" from projects associated with their sailing vessels. One such zinc provides the 1.5 pound counterweight for this project.

Photos by the author.

Dave Cornell, W9LD, first licensed as WNØMIG in 1952, was active through high school, college, and university. He's held call signs WØMIG, WA9NLD, K9BO, and G5WAP while on sabbatical near Warwick, UK. Dave held a Fulbright lectureship in Zimbabwe, where he experienced pileups from the other end as K9BO/Z2. He switched to vanity call sign W9LD, partly to remove the fragrance associated with call sign K9BO. Dave enjoys building antennas, station accessories, and keeping in touch with friends. He is also a serious amateur pianist and holds a master's degree in piano performance. Dave retired from a career as a college physics teacher, and still enjoys tutoring online students in physics and trigonometry. He can be reached at 18624 Blue Ridge Drive, Lynnwood, WA 98037 or **davidcornell123@comcast.net**.

# The Flagpole Deluxe

*If you've got to hide your antenna, make this flagpole do double duty as an efficient four-band vertical.*

By Fred J. Schnell,* W6OZF

Would you believe this flagpole is really a four-band antenna, covering 40 through 10 meters? Sturdy enough to fly the flag in a stiff breeze, this trap vertical presents a good match to 50-ohm coaxial line.

The test site for the development of the antenna was the "backyard." Upon completion the antenna was moved to its permanent location and "camouflage" was added to conceal the base, shown at left.

When I moved into a mobile home park some eight years ago I was faced with an antenna problem. The manager of the park told me I could put up an antenna at night, but would have to take it down during the day and also "stay out" of the TV system. The TV problem did not bother me, but putting up an antenna at night and taking it down every day had no appeal. I first mounted a Swantenna mobile antenna on a 42-inch, chain-link fence and got along very well for several years. But I was always dreaming of something different, something with automatic band switching.

I decided to concentrate my efforts toward building a flagpole antenna, with no bulges like the trap verticals. It had to cover 40 through 10 meters and be able to fly a 4 × 6-foot flag. I had room for neither radials nor a large copper screen, so only a ground rod could be used — a compromise. Faced with these constraints, I proceeded with the design of my flagpole.

When finished, the antenna is an aluminum tubing vertical with slim, home-built traps, covered by standard PVC pipe, and topped with a toilet-tank ball. The PVC pipe has nothing to do with the antenna itself; it is the flagpole. The base is small and simple, thus easily concealed by brickwork, rocks, or flowers.

**Table 1**

1 - Insulator block, 1 x 1-1/2 x 3/8" (25 x 38 x 10 mm) (formica, Plexiglas, plastic, etc.). 1 - Length aluminum tubing, 1-1/4 x 0.058 x 144" (3.66 m) 6061-T5.
1 - Length alumimum tubing 1-1/8 x 0.058 x 144" (3.66 m) 6061-T5.
1 - Length aluminum tubing 1-1/8 x 0.125 x 36" (914 mm) 6061-T5.
1 - Piece of aluminum rod 1 x 1" (round).
1 - Fiberglass round rod 1 x 24" (25 x 305 mm).
4 - 2" U bolts.
5 - 25-pF Centralab transmitting capacitors, 850S series.
1 - Piece of construction aluminum channel stock, 3/4 X 1-3/4 x 14" (19 x 44 x 356 mm), 1/8" wall thickness.
1 - Piece of construction aluminum T-bar stock, 3 x 1-1/2 x 14" (76 x 38 x 356 mm), 1/8" wall thickness.
8 - 1-1/4" hose clamps.
1 - 2-foot length (0.61 m) 1-1/4" EMT conduit. 1 - 3-foot length (0.91 m) 1-1/4" EMT conduit. 1 - 12-foot length (3.66 m) of 2" PVC schedule 40 water pipe.
1 - 2"- to- 2"PVC coupling. 1 - 1-1/2" PVC cap.
1 - 12-foot length (3.66 m) of 1-1/2" PVC schedule 40 water pipe.
1 - PVC reducer, 2" to 1.-1/2". 1 - Tank ball or Eagle 5".
1 - 10-foot length (3 m) copper water pipe (for ground, if required).
1 - Piece of copper or brass foil, 0.005 x 1 x 12" (0.13 x 25 x 305 mm); this may be sal vaged from an old transformer or shim stock, or may be purchased from auto supply or metal suppliers.
2 - Insulating sleeves made from plastic tubing 1-1/8" inside diameter, 1/2" wall thickness. These can be made from two PVC 1-1/4-to-3/4"bushings. These are obtainable wherever PVC pipe is sold. The bushings will have to be reamed very slightly to have a press fit over the 1-1/8 x 0.125" aluminum tubing. Either file or turn the flange off on a lathe.

## Problem Solving

A suitable ground system and skinny traps were two areas requiring some head scratching. I settled on a 10-foot (3.05 m) length of copper water pipe for the ground rod. Installation of this pipe was accomplished by soldering a hose fitting on one end of the pipe, connecting a garden hose to it, and turning the water on full force. You'll have to stand on a step ladder to do this but the method works beautifully. I have another ground rod under the mobile home, and the ground system consists of the two ground rods and also the "skin" of the home.

After much experimenting I finally settled on putting the capacitors inside the trap forms; I selected capacitors which would keep the total height of the vertical under 19 feet. It topped out at 17 feet, 9-3/4 inches (5.43 m). The capacitors used are the Centralab 850S series doorknob transmitting type. I had used them before when I built a five-band dipole which has been described in *The ARRL Antenna Book*. The first traps were made of formica tubing reinforced with Lexan rod, but I recommend fiberglass rod for the added strength and lighter weight.

## Learn to Be a Pack Rat

If you start from scratch, you will have to visit several stores in order to obtain all the materials. Try hardware, metal, building, and lumbar companies.

The aluminum tubing is 6061-T6 or equivalent, 0.058-inch wall thickness. The T- and U-bar stock is construction aluminum pieces and is available from most metal supply companies. Auto supply and discount stores are excellent sources for U bolts and clamps. The 1-1/ 4-inch conduit may be purchased from an electrician, but since you will only need 2 feet of this they will probably give you that much from their scrap heap. Assemble all the materials before you start since nothing is more exasperating than beginning a project and then not being able to readily get all the parts. **Table 1** is a complete list of the necessary parts.

## Getting Started - Preliminary Constuction

**Figure 1** shows details of the base mounting assembly. Drill and countersink a hole to accommodate a no. 8-32 × 9/16 flat-head screw. Drill and tap the insulator block and mount it inside the U channel. Drill eight 3/8-in. (10 mm) holes in the T bar as shown; the 2-in. U bolts will

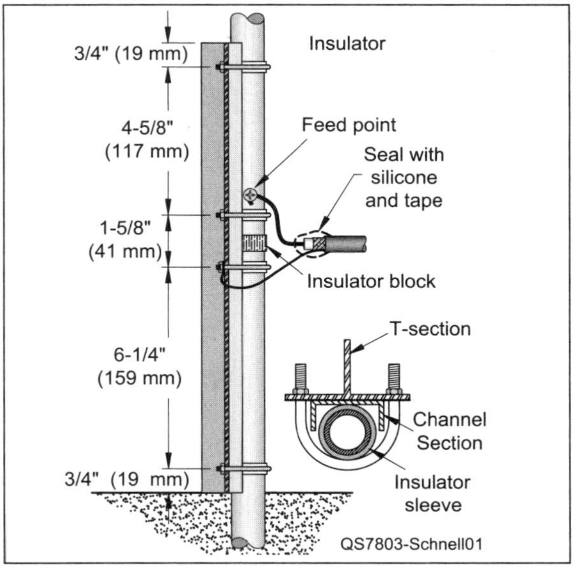


**Figure 1 — U- and T-bar stock details for the base assembly.**

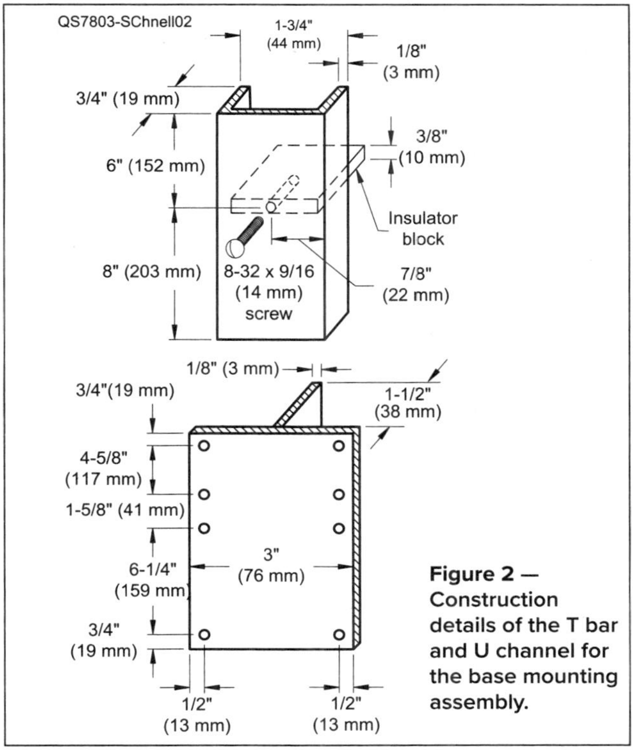


**Figure 2 — Construction details of the T bar and U channel for the base mounting assembly.**

straddle the U channel. Assemble the U and T bars per **Figure 2**, leaving the U bolts loose bars per Figure 2, leaving the U bolts loose.

Fit the insulator sleeves on the 36-in. length of 1-1/8-in., 0.125-in. wall tubing (Figure 2). These insulators should fit snugly and the U bolts which clamp them to the base assembly will hold them tight. Drill and tap the tubing for a no. 10 screw, ¼ in. (6 mm) above the top insulator; this will be the feed point.

## Preliminary Construction - Traps

Refer to **Figure 3.** Cut one 7-1/2-in. (191-mm) and two 8-in. (203-mm) lengths of 1-in. fiberglass rod, and through all three pieces drill a 1/8-in. (3-mm) hole lengthwise. In one end of the 7-1/2 in. rod, drill a 3/4-in. (19-mm) hole 3/4 in. deep. In the other two pieces, drill a hole in one end 1-1/2 in.

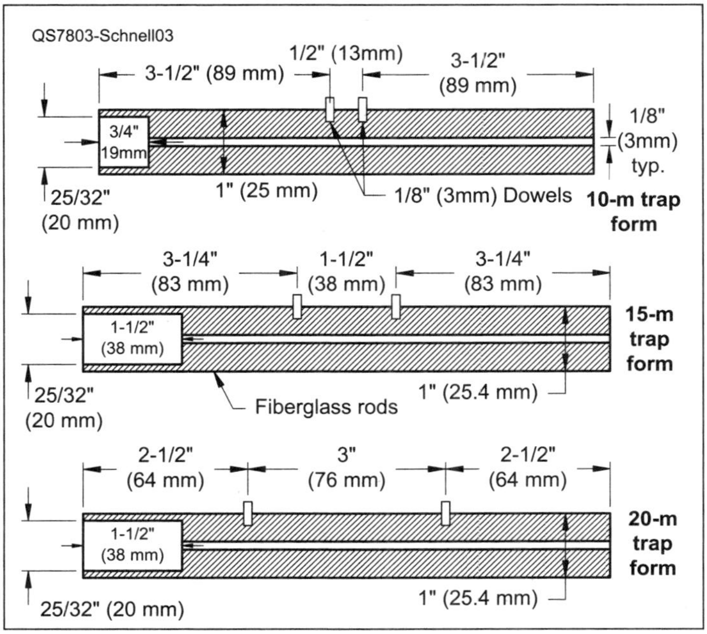


**Figure 3 — Cross-sectional views of the three traps.**

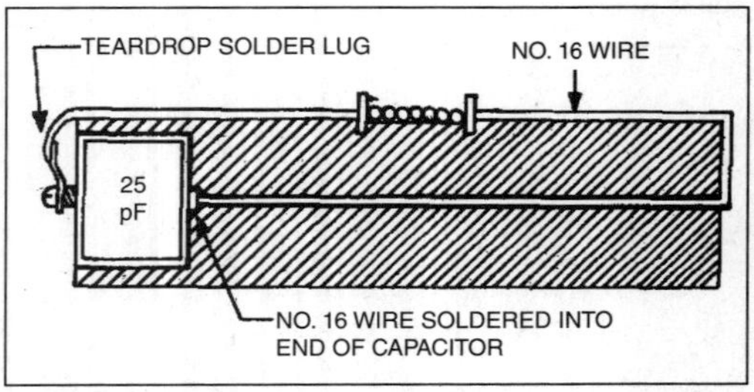


**Figure 4** — **10-meter trap winding detail.**

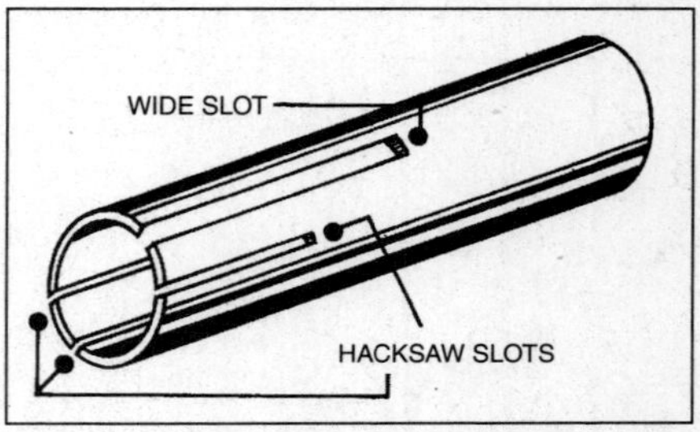


**Figure 7** — **Method of slotting the tubing for use as bushings.**

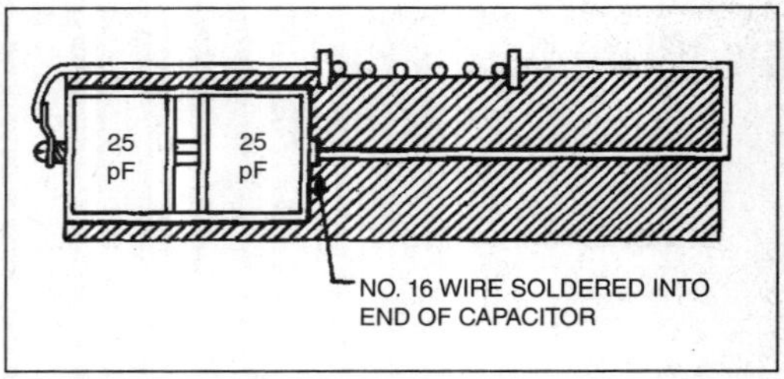


**Figure 5** — **15- and 20-meter trap winding details.**

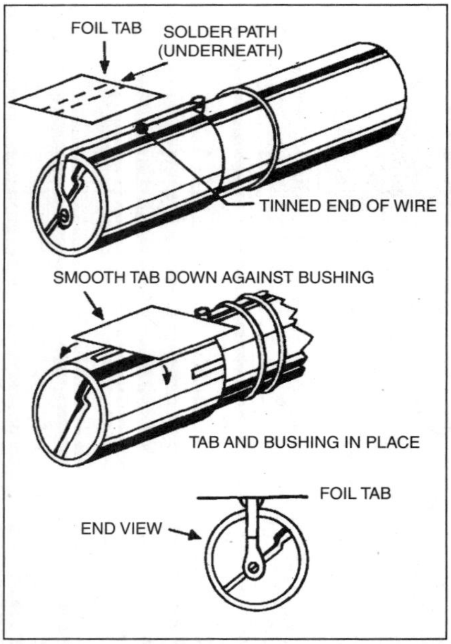


**Figure 6** — **Further details of trap construction, showing the foil tab and bushings.**

(38 mm) deep. I used masonry drill bits for these holes.

In the 7-1/2-in. rod, drill two 1/8-in. holes (3/16 in. deep) for pegs, as indicated in Figure 3. Insert 1/8-in. pegs in these holes. These pegs will be used to hold the windings in place.

Referring to **Figure 4**, wind the 10-meter trap. Begin by soldering one end of a suitable length of no. 16 wire (Formvar or Thermaleze) to one end of a 25-pF capacitor. Push the wire through the rod from the large hole end and push the capacitor all the way in, then pull the wire snugly over the end of the rod to the nearest peg and wind five turns, close spaced. Coming off the next peg, dress the wire to the capacitor. and solder to it by means of a no. 6 solder lug. This completes assembly of the 10-meter trap.

Following the same procedure, wind the 15- and 20-meter traps. Note that these traps are space wound and that the last turn is spaced a little more than the others; Figure 4 and **Figure 5** illustrate. The 15-meter trap has 13 turns spaced to a length of 1-3/8 in. (35 mm) and the 20-meter trap has 23 turns spaced to a length of 2-3/8 in. (60 mm), with the last turn of each spaced slightly more than the rest. This aids in later tuning of the traps.

Referring to Figure 5, scrape the enamel off both ends of each winding from the pegs to the ends of the fiberglass rods, and carefully tin each one with a hot soldering iron. Cut six foil tabs, each 3/4 × 1-1/2 in. (19 × 38 mm) and run a solder-path lengthwise along each tab, on one side only. Now solder these tabs onto the previously tinned wires (Figure 5).

Cut six lengths of the 1-1 /8-in. tubing, each six in. (152 mm) long. In one end only of two of the tubes cut a slot 3-1/2 in. (89 mm) long, using two hacksaw blades together in the hacksaw (**Figure 7**). This will make a slot about 1/8 in. wide. On the other four tubes, make this slot 3 in. (76 mm) long. Remove one hacksaw blade and cut three more slots in the same end as the wide slot, spacing them 90 degrees apart. Clean and deburr all cuts. These slots permit tight compression when the hose clamps are applied.

Carefully push these bushings (the slotted tubes) into place on the trap forms, up against the pegs; the longest slotted bushings go onto the 10-meter trap. Then gently press the foil tabs down against the tubing. The bushings must be in place for the following operation (**Figure 8**).

The traps should be grid dipped and adjusted for resonance at the following frequencies: 28.0 MHz, 20.5 MHz, and 14.0 MHz. Do not couple too tightly to the traps as the grid dipper can be "pulled," resulting in erroneous readings. Adjust the traps by carefully spreading or compressing the last turn on each trap.

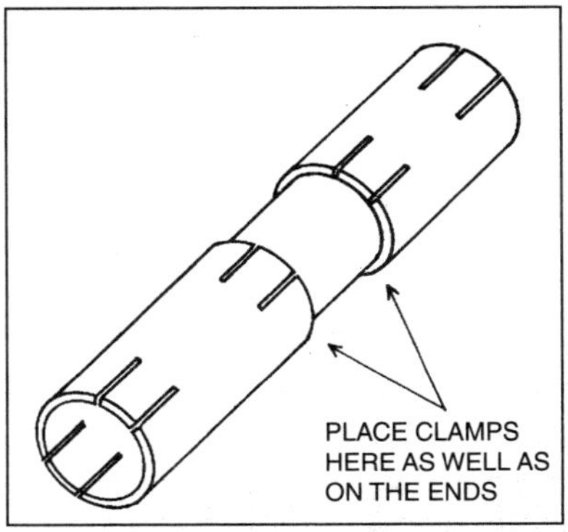


**Figure 8** — Bushings mounted in place on the 20-meter trap.

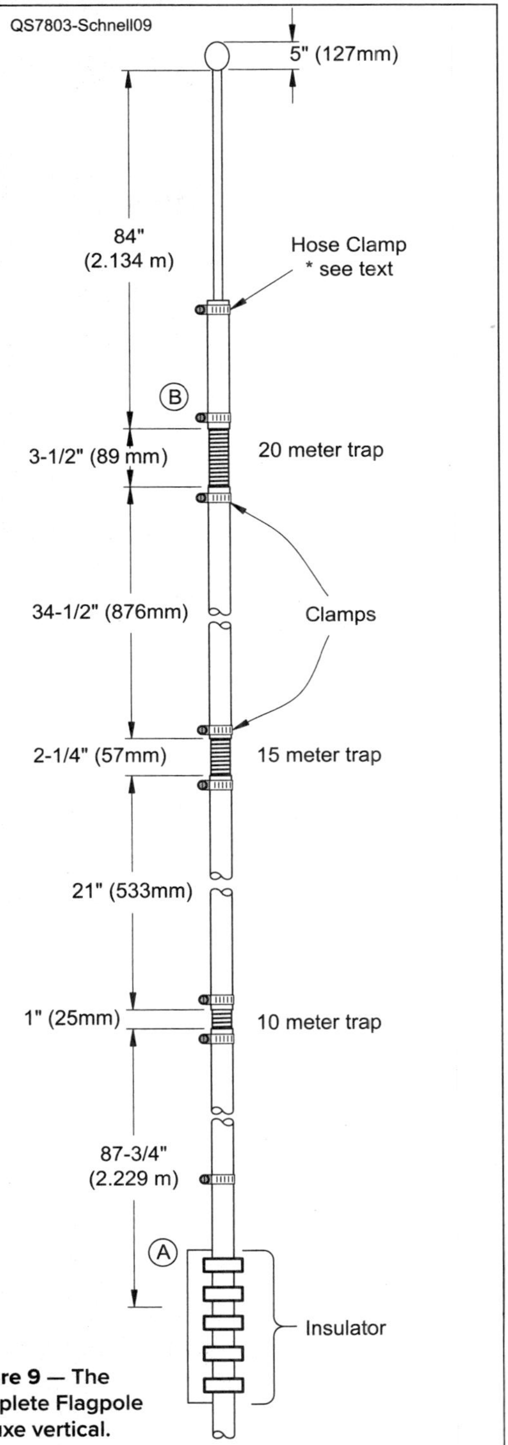


**Figure 9** — The complete Flagpole Deluxe vertical.

## Construction of Adjustable Sections

Refer to **Figure 9** and cut the 12-foot (3.66-m) length of 1-1/4-in. tubing into four lengths, as follows: 5 ft, 6 in. (1.68 m); 1 ft, 9 in. (0.53 m);* 2 ft, 10-1/2 in. (0.88 m);* 1 ft, 10-1/2 in. (0.57 m). The starred (*) items may be cut in half and a bushing inserted to permit adjustment of the 15- and 20-meter bands (Figure 8). If you choose this method, four more hose clamps will be required.

The tubes just cut are all slotted and deburred on both ends, inside and out. If this deburring is not done the aluminum may sieze or gall and it then becomes difficult (if not impossible) to separate or adjust. A stainless-steel hose clamp is placed over each slotted end of each tube.

Cut and deburr a 6-foot (l .8 m) length of the 1-1/8-inch tubing and insert the 1-inch piece of round aluminum rod, which should be drilled and tapped for 1/4-20 thread all the way through. Secure this in the end of the tubing by drilling two no. 36 holes and tapping for 6-32 screws.

## Putting It Together

Figure 9 provides an overall picture of assembly of the vertical. Nothing is critical except that care must be exercised when attaching the traps to the 1-1/4-inch tubing sections. Start by spreading *one* of the slots in the tubing ends to expand the diameter slightly. The foil on each trap must be started smoothly between the trap bushing and the tubing sections. This is made easier by carefully pushing the foil tightly against the bushing before sliding the bushing onto the trap itself.

Start by sliding the 21-inch, 15-meter tubing section onto the capacitor end of the 10-meter trap, up against the peg, and tighten the hose clamp securely. In the same manner, clamp the 34-1/2-inch, 20-meter section onto the capacitor end of the 20-meter trap. Tighten all clamps securely, then slide the 15- and 20-meter sections onto the bottoms of their respective traps. See Figure 9. Next, clamp the 6-foot length of 1-1/4-inch tubing to the bottom of the 10-meter trap, the 36-inch length of 1-1/8-inch tubing into the last tubing just installed and adjust the total length to 87-3/4 inches (2.23 m) from the 10-meter trap to the end of the tubing.

Install the ball on the end of the 6-foot length of 1-1/8-inch tubing, which has the aluminum plug in it, using a 2-inch length of 1/4-20 threaded stock. This can be cut from a 1/4-20 bolt. Install this section of tubing into the end of the tubing on the 20-meter trap and adjust to the dimension shown in Figure 9. Recheck all clamps for tightness at this time.

Complete base-mounting assembly in place.

## Testing and Adjustment

If your situation is like the author's, you will be forced to test and tune the antenna somewhere other than its eventual "camouflaged" location. Do your tuning in the final position if it is at all possible, since the effects of surrounding objects can undo an entire afternoon's work. Set the 3-foot length of 1-1/4-inch EMT conduit in concrete next to your ground rod, leaving 7-3/4 inch (197 mm) above the concrete. Mount the antenna assembly on this conduit (Figure 1), and connect a ground strap from the ground rod to the base assembly.

Once everything is in place and tightened, grid dip to determine the resonant frequencies of the vertical on each band. When making adjustments, remember that any change made on one band affects all *lower* frequency bands. Adjust the 10-meter section first, and measure the SWR by attaching the coax and applying very low power to the antenna through an appropriate SWR indicator. Continue with the other bands; very little trimming or adding should be required. If adding length is necessary, the alternate method described earlier of cutting a section of tubing in half and adding a sleeve is recommended.

After the 40-meter section is adjusted, remove the hose clamp and secure the tubing with four no. 6 sheet-metal screws. This is necessary because the 1-1/2-inch PVC pipe will not clear the hose clamp. After all adjustments have been made, SWR should not exceed 2:1 at any band edge.

## Camouflaging

The time has come to turn the vertical into a flagpole, or vice versa. Lay the entire antenna (including the base mounting assembly) flat on the ground, with the 2-inch PVC alongside it. Carefully measure the distance from the top of the base assembly (Point A, Figure 9) to the top of the 20-meter trap (Point B) and cut the PVC to length. Install the 2-inch to 1-1/2-inch reducer assembly with PVC cement. Next measure the distance from the ridge inside the reducer to the top of the tubing of the 40-meter section; make sure you have the end of the 2-inch PVC even with the top of the base assembly. Cut the 1-1/2-inch PVC to length and cement it into the reducer.

Drill a hole in the PVC cap to clear the 1/4-20 stud in the end of the 40-meter section. Remove the base assembly and slide the antenna through the PVC pipe until the stud comes out the top; slip the cap onto the stud and screw the top ball on tight. Then slide the antenna back until the cap slides all the way on in place on the end of the 1-1/2-inch PVC. Do *not* cement the cap in place. One word of caution — make certain all clamps are tight before sliding the antenna back and forth in the PVC, to prevent any change in dimensions. The end of the 2-inch PVC should now be even with the base mounting assembly when the antenna is reattached.

This completes construction of the Flagpole Deluxe. The author uses a small pulley and rope for flying a flag from his vertical, and comments from neighbors indicate full approval of the obviously handsome flagpole. Results radio-wise have been rewarding, and it is not even necessary to stop operating when it rains! Finally, the author gratefully acknowledges the assistance of George Rice, W6OGR, in the preparation of this article.

# CHAPTER 2

# Loop and Slot Antennas

# An Antenna Idea for Antenna Restricted Communities

*Dress up your garden and neighborhood with antennas that look like lawn sculptures.*

Cristian Paun, WV6N

For a number of years I used various mobile antennas while operating from home on all bands between 3.5 and 30 MHz. My score of 264,922 in the 2010 CQ World Wide DX contest using low power (272 out of 14,765 entries) shows just how good such an antenna can be, if used in a permanent installation with appropriate ground radials.[1]

Still, I wanted something that would work even better on the lower HF bands and at the same time be unobtrusive to my neighbors. What antenna could possibly come close to the performance of a full size vertical and not be as visible?

## The Small Loop Antenna versus the Competition

The small loop antenna, often called a magnetic loop, has been around for a long time. While some have had good luck with such an antenna, many others have been disappointed. I suspect they suffered from losses in loop conductors and in their interconnection method that can make such antennas much less efficient than expected. I recommend that the reader start with some useful references on this topic.[2-4]

The following will attempt to show that a properly constructed small loop antenna can perform almost as well as a properly installed elevated vertical. Since different readers will have different available space and other requirements, I have not attempted to provide detailed construction information. However, many of my design details are available on the *QST*-in-Depth website for those who are interested.[5]

### An Elevated Vertical Monopole

**Figure 1** is the elevation pattern of a 40 meter elevated vertical monopole antenna with the base 30 feet above average ground

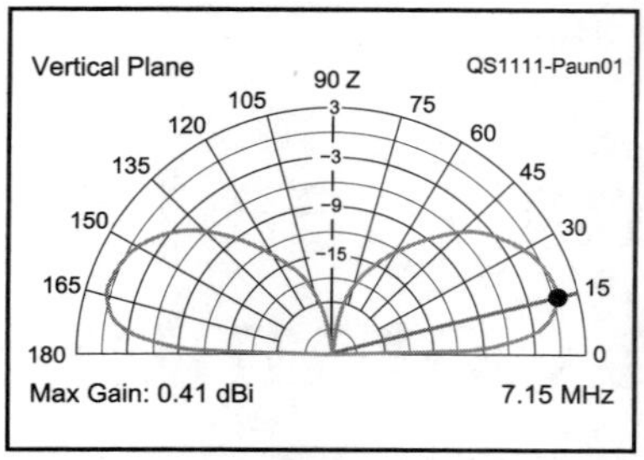


**Figure 1 — The elevation pattern of the elevated ¼ wave vertical antenna 30 feet above average ground. Note the intensity of –0.41 dBi at 15° elevation. The data for all plots is from *4NEC2* simulation software.[6]**

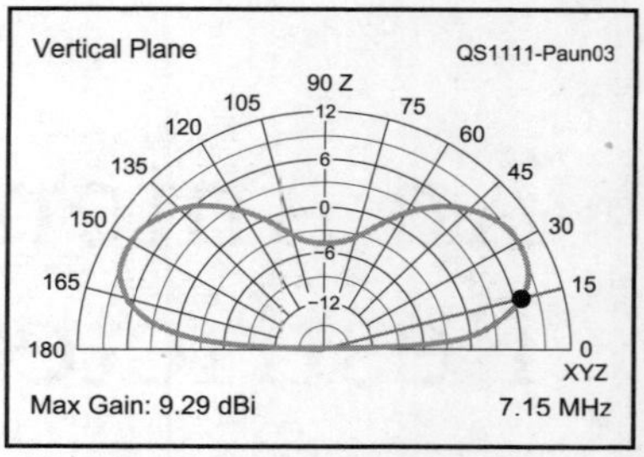


**Figure 3 — The dipole elevation pattern at the peak azimuth. Note that while the main peak is at around 30°, the intensity at 15° elevation is a respectable 6.59 dBi, more than 7 dB greater than the monopole.**

(conductivity 0.005 S/m, dielectric constant 13). At 15° elevation, a good yardstick for DX operation, the signal intensity from this vertical is about –0.41 dBi. While that doesn't sound too great, it is a lot higher than what a short mobile antenna can do. The azimuth pattern is omnidirectional, as one might expect.

Unless fed with a tuner, or traps or other schemes are employed, in its simplest form it is a single band affair.

## The Horizontal Dipole at a Height of ½ Wavelength

A dipole at this height has a significant performance advantage over a monopole due to both the ground reflection and the restricted azimuth pattern that focus its radiation in the broadside direction. The azimuth pattern at 15° elevation is shown in **Figure 2** and the elevation at the azimuth peak is shown in **Figure 3**. Note that while the main peak is at around 30°, the intensity at 15° elevation is a respectable 6.8 dBi, 7 dB greater in its preferred direction than from the monopole.

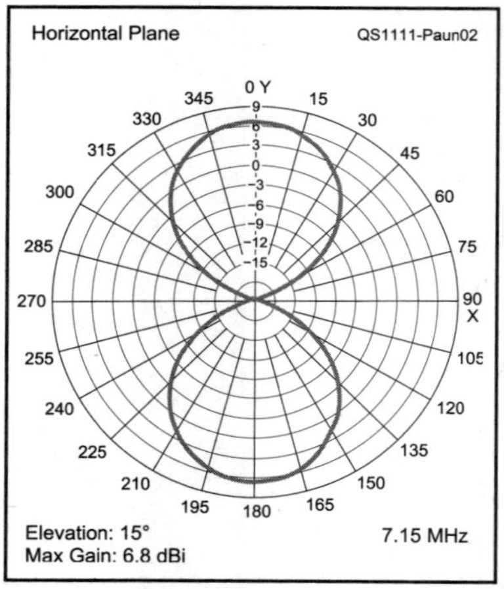


**Figure 2 — The half-wave-high dipole azimuth pattern on 7.15 MHz at an elevation angle of 15°.**

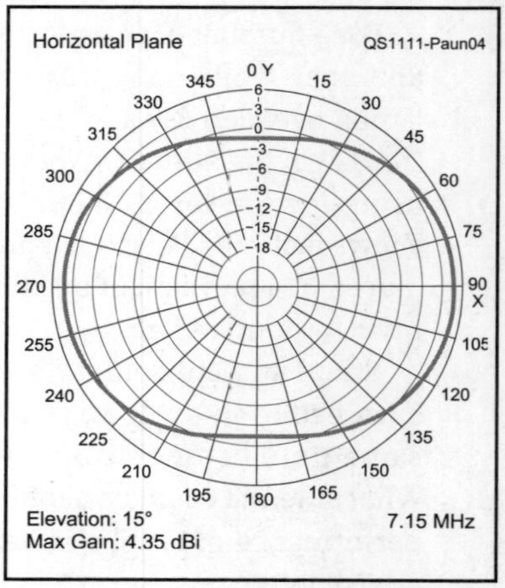


**Figure 4 — The small single turn loop azimuth pattern on 7.15 MHz at an eleva-tion angle of 15°. Note that the maximum radiation is in the plane of the loop perpen-dicular to the pattern of the more familiar large loop such as in the quad antenna.**

The dipole's disadvantages are that for the lower bands it is hard to rotate. To get worldwide coverage, it takes three of them, although two at right angles can come reasonably close. Each requires a half wave support pole on each end, although at least one could be shared. While traps can be used to operate on multiple bands, a coax fed dipole is basically a single band antenna. If fed with window line and a tuner, it will operate on all bands at frequencies higher than its half wavelength. The pattern at the higher frequencies is complex, however, and it is difficult to cover all directions with just two such antennas.

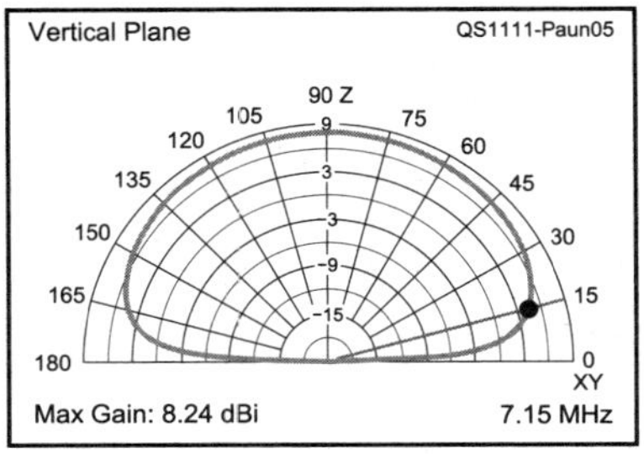


Figure 5 — The small-single-turn loop elevation pattern on 7.15 MHz at peak azimuth. Note that while the directivity is dipole-like, the loop's radiation is vertically polarized.

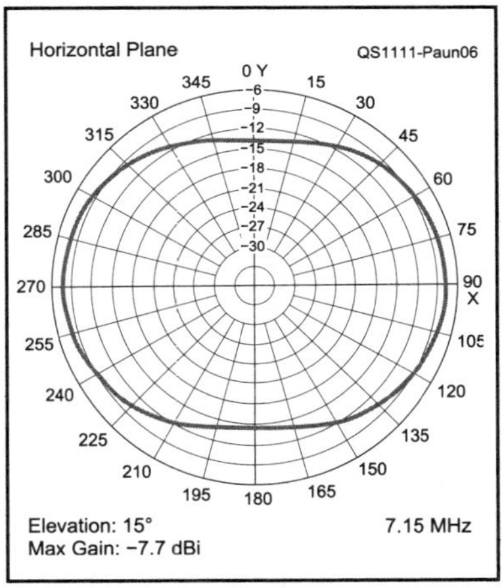


Figure 6 — The 3.3 foot diameter two turn loop azimuth pattern on 7.15 MHz at an elevation angle of 15°.

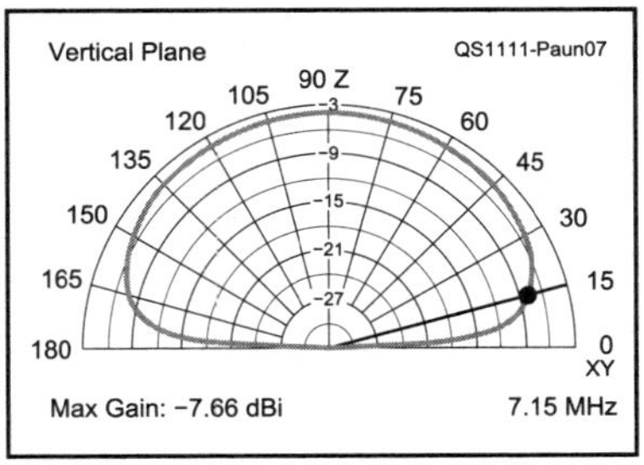


Figure 7 — The 3.3 foot diameter two turn loop elevation pattern at the peak azimuth.

### The Small Loop Antenna

An ideal small loop at a height of 10 feet has azimuth and elevation patterns as shown in **Figures 4** and **5** respectively. The improvement in performance compared to the vertical monopole is notable — 5.47 dBi at 15° elevation versus –0.41 dBi at the same angle for the monopole. That is almost exactly 1 S-unit, similar to the gain of a small Yagi over a dipole. The comparison above assumes 100% efficiency, which is not quite attainable in practice. More about that later.

The same loop at a height of 5 feet above ground has slightly less gain, 4.08 dBi at 15° — still a lot better than the vertical.

## Building the Small Loop

I wanted my loop to be tunable from 3.5 to above 11 MHz, thus covering the 80, 40, and 30 meter bands, as well as our 60 meter channels. I started with the spreadsheet from Steve Yates's website.[7]

The distributed capacitance would make tuning at 14.2 MHz rather difficult if not impossible. A 6.6 foot diameter loop has a calculated efficiency of only 20% at 3.5 MHz but that is quite a bit better than most commercial mobile antennas. The predicted (4NEC2) efficiency is significantly higher on the 7 and 10 MHz bands. All said and done, that size loop is not going to be easily hidden, especially considering the directivity of this antenna and the need to make it rotatable.

Well, since the antenna is essentially a large one-turn coil resonated by a variable capacitor, I will get the same equivalent inductance by halving the diameter but using two turns instead of one. This will require approximately the same length of copper pipe that I used to minimize loss. This is particularly important because the radiation resistance will be reduced by a factor of four in the process.[8] I took this change into account in my assessment by inserting additional series resistance into AA5TB's spreadsheet calculator. Azimuth and elevation patterns are shown in **Figures 6** and **7**, respectively.

An additional price to pay for the convenience is the added distributed capacitance, which could make tuning difficult at the upper limit of the frequency range. Enough design margin can take care of that, though. With the above design data, I started to build a two turn loop, about 3.3 feet in diameter, made out of off-the-shelf, commercially available 1.5 inch copper pipe. I tuned it with a motor driven 12-500 pF, 15 kV vacuum variable capacitor. The rotator is a simple TV antenna unit.

The octagonal shape was an obvious choice because the 45° fitting pieces are also readily available. The vacuum variable capacitor is controlled remotely via an electric motor with planetary reduction gear. The supporting hardware is mostly homebrew with some wooden stands that were purchased for convenience. The two turns are 5 inches apart. This antenna has been in service at WV6N since April 11, 2010.

Over the air comparison between this antenna and the vertical, with help from my friends Jim, K6JRF and Randy, WB6MMJ, showed that it delivered 1 S-unit improvement over my best mobile vertical on 75 meters. Once that orange tree grows taller, I may be able to raise this loop another 2 to 3 feet higher.

## On to Higher Bands

After a couple of months of testing this loop, it was time to build one for the higher frequency bands. I again used the AA5TB spreadsheet calculator. I wanted this version to tune from below 14 MHz all the way to past 30 MHz.

**Figure 8 — The 3 foot diameter single turn loop for 14-30 MHz. Note the coupling loop used for excitation in both loop designs.**

For good design margin, a 5-100 pF, 25 kV variable capacitor was selected. I used the same type of motor with planetary reduction gear and the same antenna rotator as the lower frequency version. The vacuum variable capacitor tuning shaft is hot with RF, so the coupling between this shaft and the motor has to be able to withstand the very high voltages predicted by AA5TB's spreadsheet. The resulting antenna is shown in **Figure 8**. This second loop tunes as intended and it can take 1500 W with the use of the 25 kV capacitor.

All the wooden parts, the stand, the frame, and the mast are home brew. I chose the octagonal shape again for convenience.

## Conclusion

Small or magnetic loops turn out to be a viable choice for those hams living in restricted areas. Depending on the efficiency of the loop, they can come close to the performance of a full size dipole at a height of ½ wavelength. They don't quite look like antennas and will outperform an elevated vertical installed at the same height and with comparable efficiency. They radiate well at low angles for DX contacts as well as at high angles for short skip communications.

The thicker the copper pipe the more efficient they will be, and the higher voltage rating of the capacitor the more power they can take. These loops are narrow band and they exhibit directivity which could be a great advantage on receive. I used to get great reports from DX stations on 20 meters with the vertical while I was barely hearing them. It had to do with the local noise being captured by the antenna and with the take-off angles at each end. Not so with the loop. They hear me well and I hear them equally well.

The designs presented here are good compromises between performance and cost. They can handle significant power, at least 1 kW for the low bands (1.5 kW if 18 kV or higher rating capacitors are used) and 1.5 kW for the upper bands. Note that with high power levels and heights potentially near people, it is particularly important that an RF safety assessment be conducted in combination with methods of keeping people far enough away to be safe.

**Notes**

[1]cqwpx.com/claimedcall.htm

[2]For a good introduction to the basics of the magnetic loop, see en.wikipedia.org/wiki/Magnetic_loop.

[3]Practical information regarding the building of the magnetic loops is available at standpipe.com/w2bri/.

[4]This article makes extensive use of the EXCEL calculator that Stephen Yates, AA5TB, makes available on aa5tb.com/loop.html.

[5]arrl.org/qst-in-depth

[6]4NEC2 is available to amateurs by Arie Voors without cost at home.ict.nl/~arivoors/Home.htm.

[7]See Note 4.

[8]The radiation resistance of a small multiturn loop can be shown to be RR = 31,171 × (n × A/l2) 2. See, for example, J. Kraus, Antennas (New York: McGraw Hill, 1950), p 167.

Photos by the author.

ARRL member and Amateur Extra class operator Cristian Paun, WV6N, has been interested in ham radio for many years, He started as a shortwave listener in 1983 and then became licensed as YO3FMY. He moved to Australia and there became VK3MS (still an active call sign). He now lives and works in California and operates as WV6N.

He holds an MS degree in Electrical Engineering. While living in Australia he was employed by Codan and Winradio. He now works as an RF engineer, designing radios and antennas for a large corporation. Cristian holds a number of US and European patents for antennas. You can reach Cristian at 904 Aster Ln, Lompoc, CA 93436-3241 or at **wv6n@arrl.net**.

# Nested Loop Antennas

*This low-cost five band loop array blends into the background.*

G. Scott Davis, N3FJP

This stealthy nested loop is almost invisible among the trees.

This multi-band nested loop antenna array replaces my tribander Yagi, which is only up 20 feet. Inspired by suggestions from Bill Wisel, K3KEI, I first tried a full wave 20 meter band square loop antenna. On the air comparisons with my low Yagi confirmed instantly that this design was a hands-down winner for working both local and distant stations.

I replaced that mono-band loop with a nested loop array for the 20, 17, 15, 12, and 10 meter bands. The antenna blends into the surroundings, so I needed the morning sun shining directly on it to snap the lead photo.

This became a nice father-son project with my son Brad, KB3MNE. Here's how we built the antenna.

## Construction

We constructed the square loops shown in **Figure 1** according to the dimensions in **Table 1**. The loops hang from a tree limb in the vertical plane. Because I feed them from the bottom corners, the loops radiate horizontal polarization.

Calculate the perimeter size, P, of each loop by dividing the frequency in MHz into 1005 feet. Table 1 shows the loop dimensions. Start with the 20 meter loop, the largest loop. Cut a

**Table 1**

**Nested Loop Antenna Dimensions; 75 W Matching Section Cable has 0.66 Velocity Factor**

| *Band* | *Frequency, MHz* | *Perimeter, P, ft* | *Side, S, ft* | *Cross Support, ft* | *Distance From Center, D, ft* | *Length, L, of 75 W cable* |
|---|---|---|---|---|---|---|
| 20 m | 14.175 | 70.90 | 17.72 | 24.81 | 12.41 | 11.45 |
| 17 m | 18.118 | 55.47 | 13.87 | 19.41 | 9.71 | 8.96 |
| 15 m | 21.225 | 47.35 | 11.84 | 16.57 | 9.29 | 7.65 |
| 12 m | 24.940 | 40.30 | 10.07 | 14.10 | 7.05 | 6.51 |
| 10 m | 28.500 | 35.26 | 8.82 | 12.34 | 6.17 | 5.70 |

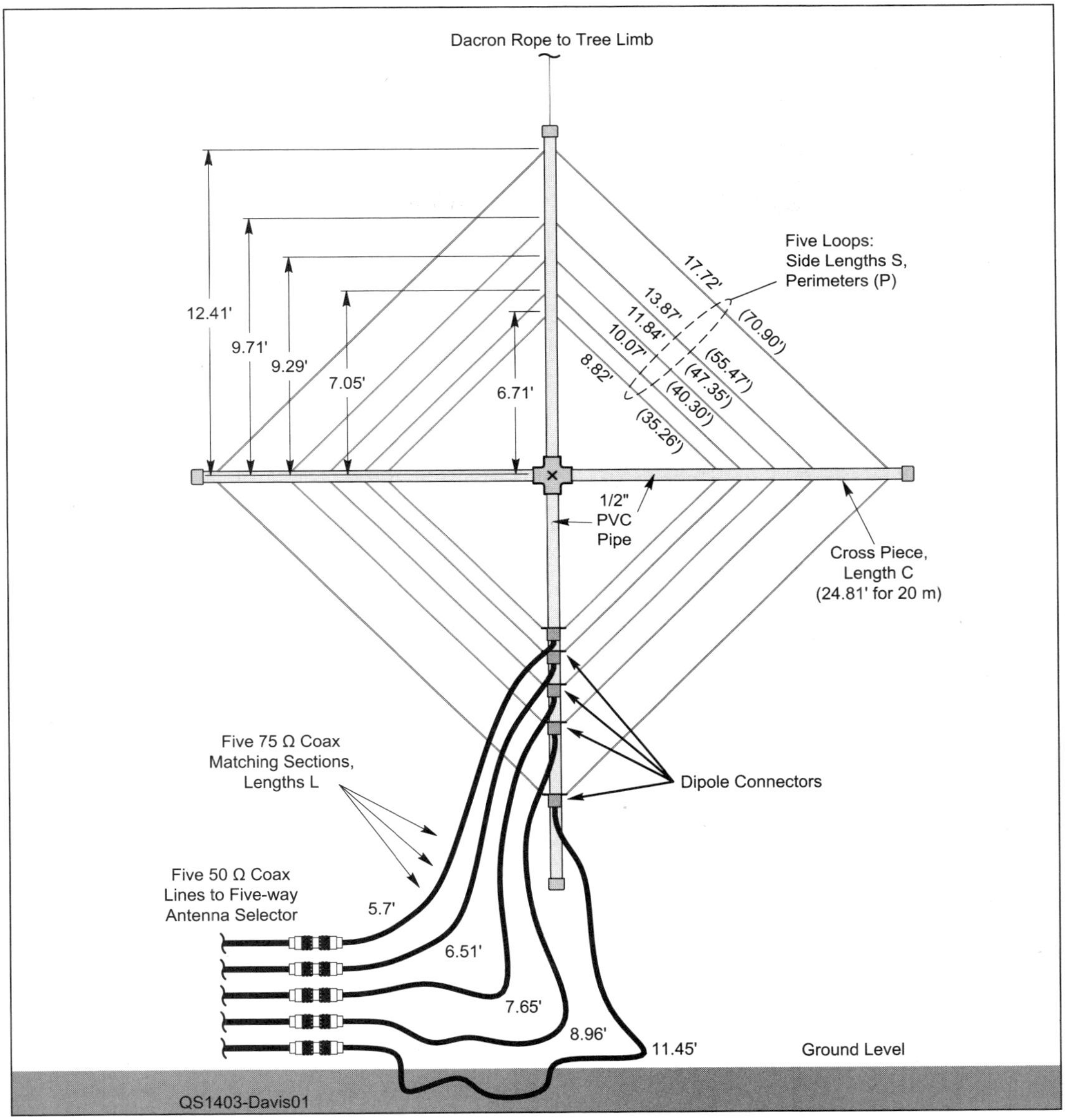


**Figure 1** — The five-band loop antenna and construction overview.

Antenna wire — #12 AWG black insulated stranded copper wire, 250 feet (**homedepot.com**).
Coax cable — 75 Ω, 41 feet (RG11A/U or RG59B/U **thewireman.com**).
CQ dipole center hang, SO239 connector, 5 units (Part #801 **thewireman.com**).
Double barrel female SO-239 connectors, 5 units
PL259 coax connectors, 10 units.
50 feet of ½" PVC tubing (four 12.5" sections).
PVC 4 way connector, ½".
PVC ½" end caps, 4 units.
Wood stiffeners, 1" × 2" × 8" (optional) pressure treated wood.
Dacron UV resistant 3⁄16" rope (Part #816 **thewireman.com**).
Brass wood screws, #6 x 1½", and PVC cement.

**Figure 2 —The five SO239 dipole connectors attach the loops to 75 Ω coax quarter wave matching sections. Brad, KB3MNE, points to the 17 meter connector.**

70.90 foot segment of wire for the loop perimeter length, P. Divide the perimeter length by 4 to determine the length, S, of each side. The PVC pipe horizontal cross support spans between opposite loop corners, and is 1.41 times the largest loop side length, S. We used brass screws to hold the PVC together. Lay your antenna wire out on the structure, temporarily taping it to find where the wire should pass through the PVC pipe. Drill holes through the pipe for the loop wire. After you run the wire through the holes, wrap a bit of electrical tape on each side of the wire next to the pipe to keep the wire from sliding and to give the pipe additional support. Next, cut and thread the wires for the rest of the bands.

The horizontal PVC pipe sagged and bowed significantly even though the antenna wires help support the horizontal PVC pipe pieces. We remedied the sagging by taping lengths of 1 × 2 inch pressure treated wood to the horizontal arms. At the bottom, rather than sliding the wires through the PVC pipe, we put brass wood screws into the PVC to hang the dipole connectors seen in **Figure 2**.

## Matching and Feeding

Each loop antenna feed point impedance is about 100 W, so we used individual quarter wave 75 W coax matching sections to convert the impedance to 50 W. Carefully cut 75 W coax to the lengths L specified in Table 1. Those lengths, taking into account velocity factor v = 0.66, are L = 246v/FMHz feet. Connect one end of the 75 W coax to the loop using a dipole center connector, as seen in the lead photo. The other end connects to the 50 W coax feed line using a double barrel female SO-239 connector. I select the appropriate antenna by an antenna switch controller in my ham shack.

## Conclusion

My nested loop array works well on all five bands. There is little interaction between the antenna elements, and the signal reports are very encouraging. You can build a single loop for your favorite band, or nest multiple loops. You can even add 6 and 2 meters if you like. [Be sure to see the compliance tables in RF Exposure and You by Ed Hare, W1RFI. — Ed.]

---

Photos by the author.

G. Scott Davis, N3FJP, holds an Amateur Extra class license. He was first licensed in 1976 as WB3CZF. Scott shares his interest in Amateur Radio and programming with his wife Kimberly, KA3SEQ, and sons Christopher, KB3KCN, and Bradley, KB3MNE. You can reach Scott at **snkdavis@aol.com**.

# A Durable, Low-Profile 2-Meter Mobile Antenna

*A vertically polarized slot antenna designed to withstand the rigors of automotive rooftop mounting.*

John Portune, W6NBC*

Originally developed in 1938 for VHF television broadcasting, the slot antenna has not found frequent application in Amateur Radio. This design, shown mounted to a truck rooftop in the lead photo, is an exception. Being horizontal, it is naturally low-profile, yet it is vertically polarized, which is essential for mobile operation. Its omnidirectional azimuth gain is 4.83 dBi when mounted on a vehicle roof. In the same location, a quarter-wavelength vertical spike has an azimuth gain of 3.74 dBi (see **Figure 1**). The 3 dB bandwidth is greater than 5 MHz, as shown in **Figure 2**.

## Basic Theory

This antenna is a derivative of the classical slot antenna, but it is not a patch antenna. Patch antennas radiate at right angles to their metal surfaces and are therefore usually mounted vertically. Also, their surfaces are not electrically connected. In this slot antenna, the conductive surfaces are connected by two ¾-inch-diameter aluminum shorting posts. Together, the surfaces and shorting posts form a half-wavelength horizontal slot antenna.

In contrast to a conventional slot antenna, the slot and antenna surfaces are not in the same plane. The slot, formed by the aluminum shorting bars connecting the two antenna surfaces, is orthogonal to the two planes. Even though the plane of the slot is vertical, it still acts as a horizontal slot because it connects two horizontal antenna surfaces. Again, as stated above, a horizontal slot yields vertical polarization, which is essential to mobile operation. This 90-degree rotation of polarization is a fundamental difference between slots and dipoles.

## Construction

The assembly drawing is shown in **Figure 3**. Construction is straightforward and can be accomplished with hand tools. Although

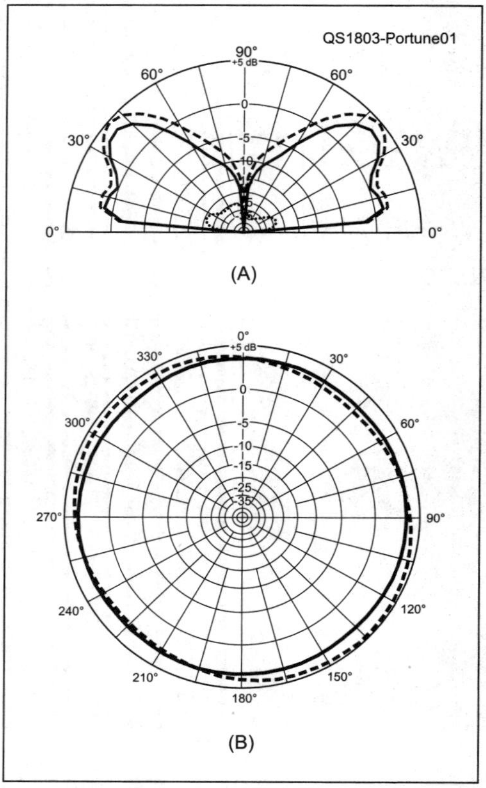


**Figure 1** — Omnidirectional azimuth gain of the slot antenna (dotted) compared with a quarter-wave spike (solid), both mounted on a vehicle roof.

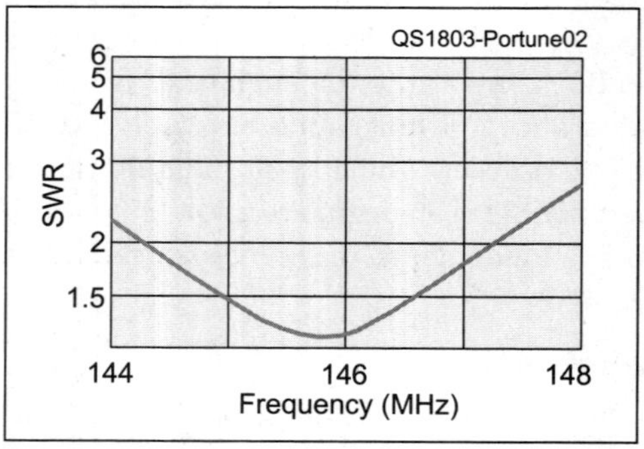


**Figure 2** — The 3 dB bandwidth of the antenna is greater than 5 MHz.

aluminum is easier to work, it is not essential; steel is quite satisfactory because conductor resistance is not a problem due to the large areas. Steel, however, must be protected from the weather. Also, the metal does not have to be solid. Perforated metal or wire cloth may be used because RF does not "see" small openings.

When forming the slots (either by drilling or cutting) for the shorting rod adjustment, clamp the top and bottom panels together so that the slots will register. The adjustment slot for the feed rod is only required in the top panel. Drilling, threading, and tapping aluminum is much easier if cutting fluid formulated for aluminum switch is used, such as Tap Magic Aluminum.

Four neodymium cup magnets mount the antenna on the vehicle. This attachment method has been extensively tested at highway speeds.

## Tuning and Matching

The match to 50 Ω (i.e., low SWR) is easily achieved with an antenna analyzer, simply by moving the feed-point fixture. Adjusting the shorting post opposite the feed point sets the operating frequency. As shown in Figure 3, the tuning range is roughly 141 – 148 MHz.

To tune the antenna, first install it in its intended operating location. Next, set the match by adjusting the position of the aluminum feed post. Finally, set the operating frequency by adjusting the position of the aluminum shorting post in its slot, diagonally opposite the feed post. Once set, the match should change little with adjustment of the frequency.

The size and shape of the vehicle will change the match and tuning. It will change even more if the antenna is used on a non-

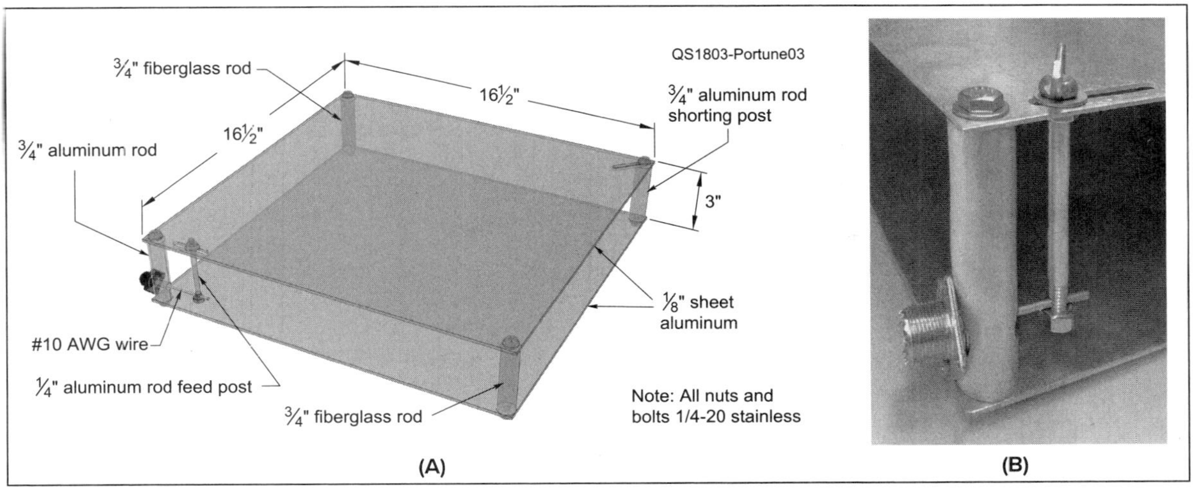


**Figure 3** — At (A) is the mechanical layout of the antenna. Note the feed point in the lower left corner: the shield of the feed line is connected to the lower plate and the inner conductor passes through the aluminum shorting rod and connects to the upper plate via a ¼-inch diameter threaded aluminum rod. At (B) is a close-up of the adjustable feed point.

metallic surface, such as a rooftop. So, if the antenna is moved from its initial setup location, it will probably need to be retuned at its new location.

## Shape Variations

The distance between the metal plates can be varied, which will affect the bandwidth, with a wider spacing having a wider bandwidth. The 3-inch spacing in the prototype is a good compromise between a low profile and a low SWR across the 2-meter band.

Similarly, the shape of the metal planes is not rigidly defined. As the slot aspect ratio changes or the rectangle is morphed into a curved shape, the radiation pattern will change, but the operating characteristics will remain essentially constant. A circular shape is the most compact and is omnidirectional, but the square version is the easiest to construct. The trade-off is yours to make.

Merely as examples, **Figure 4** shows two other shapes created during prototyping. Aesthetically, the boomerang shape (see

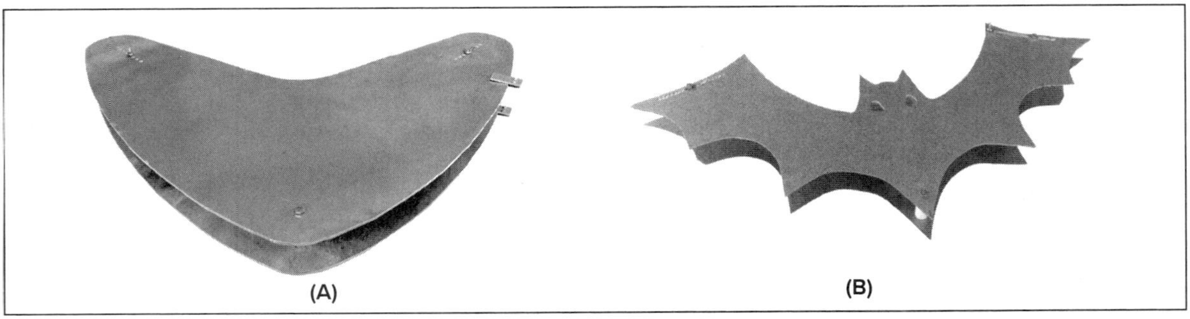


**Figure 4** — Alternative shapes are also possible. At (A), a boomerang shape is aesthetically more pleasing than the square, but requires more fabrication effort. At (B), the novelty bat-shaped antenna is always a fun attention-getter.

Figure 4A) is more attractive than a simple square. The bat-shaped version (see Figure 4B) was created purely for fun and is entirely functional. It is popular in parking lots and at radio club meetings and rallies.

## Conclusion

This orthogonal-plane slot antenna is ideal for low-profile mobile operation on RVs, trucks, and SUVs, where durability is required. However, it may also find application on military vehicles, owing to its ruggedness and an unconventional appearance that is not readily seen as an antenna. Its stealthy profile also makes it ideal for fixed rooftop use, especially metal roofs, due to the fact that no ground plane is needed.

---

Photos by the author.

John Portune, W6NBC, is an ARRL member and frequent contributor to *QST*. He has been licensed for 52 years and has held an Amateur Extra-class license since 1972. John has a BS in physics and also holds FCC Commercial General Radiotelephone Operator and FCC Radiotelegraph licenses. He retired as a broadcast television engineer and technical instructor at KNBC in Burbank, and then from Sony Electronics in San Jose, California. You can reach John via e-mail at jportune@aol.com or through his website at **w6nbc.com.**

### Materials List

(2) 16½ × 16½ × ⅛ inch aluminum sheets (can be as thin as 0.09 inches or substitute perforated metal or heavy wire cloth)

(2) ¾ × 3 inch aluminum shorting rods tapped for ¼-20 bolts at both ends

(2) ¾ × 3 inch fiberglass or acrylic spacing rods tapped for ¼-20 bolts at both ends

(1) ¼ × 3 inch aluminum connecting rod threaded for ¼-20 nuts at both ends, one end cross drilled with a ⅛-inch hole, 5⁄16 inch from the end

(8) ¼-20 × ¾ inch stainless bolts and washers

(4) ¼-20 stainless nuts

(4) 1¼-inch neodymium cup magnet mounts with holes for mounting

(4) stainless mounting screws and nuts for cup magnets

(1) chassis-mount connector of your choice with mounting hardware

# A Stealth Rooftop Antenna

*This 2-meter antenna is only 12 inches tall and can slip over a vent pipe on your roof.*

John Portune, W6NBC

Ham antennas are eyesores to many of our neighbors, and homeowners associations often forbid them. But vent pipes, fireplace flues, and air-conditioning ducts — which are all just as visible as antennas — don't seem to be bothersome. If a ham radio antenna can be made to look like a normal rooftop dweller, no one will ever notice it.

I constructed a tiny 2-meter staggered-slot rooftop antenna that's only 12 inches tall and looks like a vent pipe (see **Figure 1**). This antenna has a gain of roughly only one S-unit below a **J** pole, while being much smaller.

## Materials

This small wonder is made from 12 inches of PVC pipe that's 3 or 4 inches in diameter and heavy-duty 2-inch-wide aluminum foil tape (see **Figure 2**). You may want to add additional length to the bottom end of the pipe for mounting or for cutting an angle to match your roof.

I used 3.6-millimeter Scotch® 3311 aluminum foil tape. Small rolls of foil tape sold by Scotch and other brands are readily available on the internet. Thinner tape or household aluminum foil and glue can also be used, but they're somewhat more difficult to apply. Copper foil tape is another alternative.

The completed vent antenna is spray-painted matte black or covered with matte self-adhesive vinyl. Mine simply sits over an existing vent pipe (not a hot one).

You'll also need masking tape, a coax pigtail (coaxial cable designed to bridge between the main coax to your radio and the antenna) — RG-58 or RG-8X (Mini-8) are satisfactory — ring terminals for 6-32 screws, and brass or stainless-steel hardware.

## Building the Antenna

Begin by wrapping masking tape around the pipe 12 inches from the top to mark the bottom

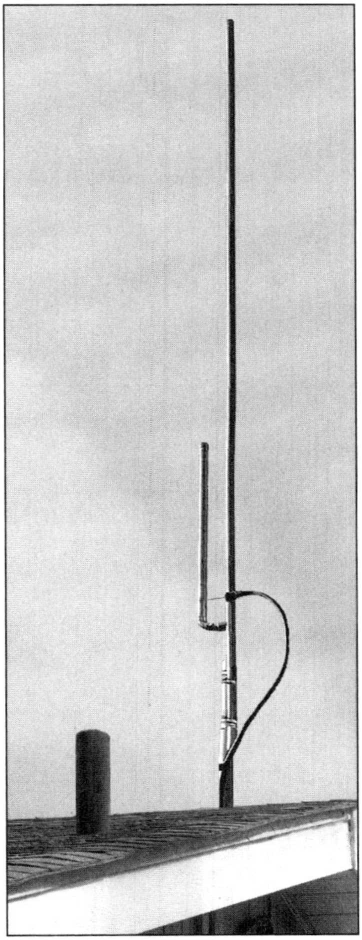

**Figure 1** — My homebrew J pole next to the vent pipe rooftop antenna.

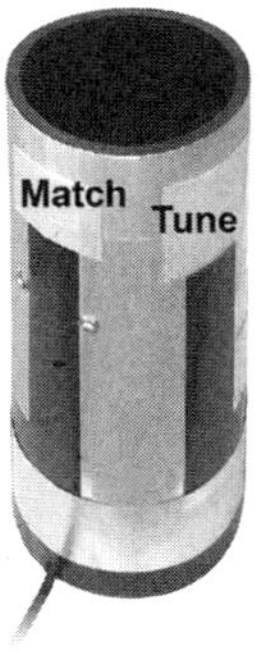


**Figure 2** — This stealth rooftop antenna is made using PVC pipe and heavy-duty 2-inch-wide foil tape, with two, 9/64-inch holes drilled near the edge of the staggered slots for attaching a coax pigtail from the inside of the pipe. Note the highlighted shorting tapes.

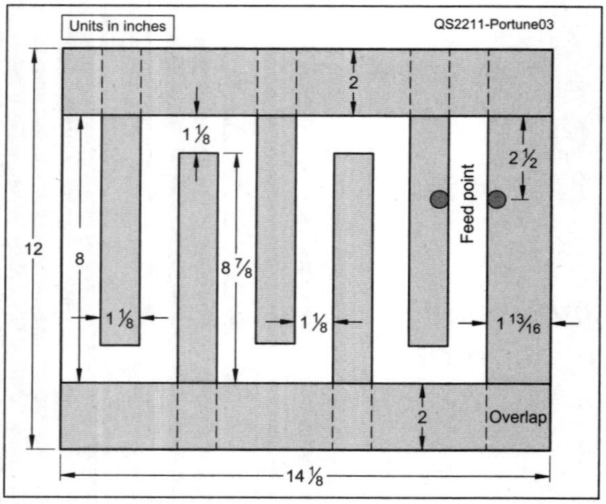


**Figure 3 — A flat view of the antenna design shows the foil tape segments and locations for the 4-inch version. The feed point is between the two darker shaded dots on the right side. dots. The left and right edges of this diagram overlap when wrapped around the pipe.**

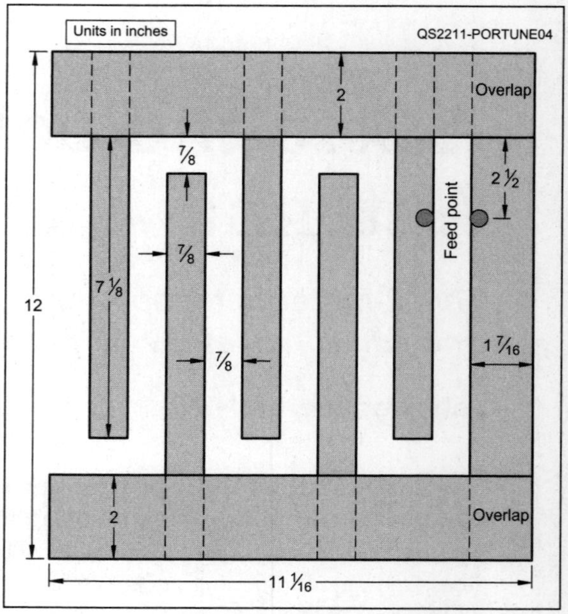


**Figure 4 — A flat view of the dimensions for a 3-inch version of the stealth rooftop antenna.**

end of the antenna. Next, using the foil tape, attach vertical segments that are 8⅞ inches in length, some going down from the top of the pipe and some going up from the masking tape. See **Figure 3** for the layout. Note that the vertical segments (shown in gray) alternate — one up, one down — and their ends will be overlapped by the horizontal ring segments, which wrap around the pipe and whose ends also overlap each other. This permits a good RF connection by capacitive coupling, so that the non-conductive adhesive on the tape isn't a problem. No soldering is required.

In Figure 3, the folded slot that forms the antenna is easy to see. The white serpentine area, where there's no foil tape, is a half wave in total length. With a slot antenna in this orientation, even when it's rolled around the PVC pipe, most of the radiation is vertically polarized. This is needed for a 2-meter FM base-station antenna. I used the printed markings on the PVC pipe to get the first vertical tape segment straight. Note that the right vertical segment is wider than the others and is 12 inches long. The right edge in Figure 3 forms the left edge of the leftmost slot section, thus ending the slot near the top of the antenna. Complete the foil tape application by adding 2-inch-wide horizontal segments. Both of these wrap around the full diameter with an end overlap, so they should be cut an inch or more longer than the pipe diameter. Make sure all pieces of tape are smooth so they adhere well to the pipe and to each other, especially at the overlapped places.

If you prefer, the same concept can be implemented using PVC pipe of a smaller diameter. **Figure 4** provides dimensional details for a 3-inch-wide version. This smaller pipe width may be easier to find in your local hardware store. The bandwidth and gain are a little less with this pipe size, but the radiation pattern remains the same. The only difference

Figure 5 — A view of the internal feed point and coax balun.

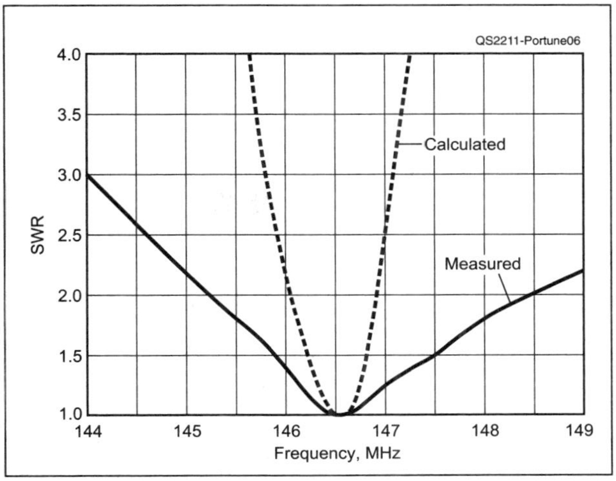


Figure 6 — SWR and bandwidth for the *EZNEC* modeled antenna (dotted curve) and as built (solid curve).

in construction is that the vertical foil tape segments and the slots are narrowed to ⅞ inch. One small downside is that attaching the coax to the internal feed point is more difficult with the smaller pipe.

## Feed Point and Balun

Make up your coax pigtail. Separate the cable's braid and the center conductor, and add ring terminals for 6-32 screws. Also apply weatherproofing. Drill two, 9⁄64-inch holes near the edge of the slots (as shown in Figure 2) for attaching the pigtail from the inside of the pipe. Use brass or stainless-steel hardware. Run the coax out through the bottom of the pipe.

If you're using RG-58 or Mini-8, allow roughly 2 feet extra in length for a four-turn coax balun, 1 inch in diameter, inside the pipe (see **Figure 5**). If this antenna will be used over an existing rooftop vent pipe, make the balun 3 inches in diameter and create a two-turn balun to slip over the vent pipe. Mix 61 ferrite beads are also suitable. To keep out bugs and rain, you may want to cover the antenna with a cap.

## Tuning and Matching

First, measure the initial resonant frequency (the tune) and the SWR (the match) with an antenna analyzer. The frequency will initially be low and the SWR will be high. The slots are longer than needed to allow for tuning and matching. To adjust the frequency, short-circuit the non-feed end of the slots with a short piece of foil tape. Using the same process, short-circuit the feed end to adjust the SWR. Note the highlighted shorting tapes in Figure 2. Completely cover any unused slot with the shorting tapes. The tapes will vary in length and position, as shown, depending on your situation. However, by proceeding in small increments, a good tune and match can always be achieved. Go slowly; the adjustments interact. If the match is best but not perfect, with no short on the fed end, move the feed point down another inch and try again.

As shown in **Figure 6**, the SWR 2:1 bandwidth is roughly 2.9 MHz, which is enough to cover the FM portion of the 2-meter band. If one compares this actual bandwidth to the lossless bandwidth of 0.8 MHz, as predicted by *EZNEC*, an estimate of radiation efficiency can be inferred. Antenna engineers generally accept that radiation efficiency is roughly the inverse of the bandwidth ratio. In this case, a 3.8 to 1 actual-to-theoretical bandwidth ratio indicates a radiation efficiency of roughly 28% or –5.5 dB. This somewhat lowered efficiency is typical of electrically small antennas. The pattern gain toward the horizon is also roughly 1 dB less than the J pole or dipole in that the wrapped-up geometry puts some energy straight up and down, leaving less for the main lobe. Field measurements confirm that the gain is down about 6 – 7 dB from a vertical half-wave dipole, and that the azimuth pattern is nearly omnidirectional. The 5.5 dB efficiency loss could be due to resistive loss in the thin foil tape and/or dielectric loss in the PVC substrate. Builders are encouraged to try better pipe material, possibly high-density polyethylene (HDPE), and to use thicker conductors. Any improvement in the efficiency will be indicated by a narrower bandwidth. But don't make it too narrow to cover all of the repeater frequencies of interest.

## Final Considerations

To complete the disguise, paint the antenna with a plastic-based paint, such as Krylon® Fusion All-In-One® spray paint. Alternatively, you can cover it with matte black self-adhesive vinyl from a local craft store or sign shop. As mentioned earlier, you may wish to cut the bottom of the pipe for your roof angle, or for a mounting clamp. You could also add a roof vent boot for extra support and waterproofing.

---

All photos by the author.

John Portune, W6NBC, was first licensed in 1965 as WB6ZCT, and he still holds the call sign MØGCK. He became an Amateur Extra-class licensee in 1972, with the call sign W6NBC, and he has his General Radiotelephone Operator License (GROL). John received a bachelor's degree in physics from Oregon State University in 1960. He's now retired but has worked as a TV broadcast television instructor for KNBC Los Angeles and Sony Broadcast, San Francisco. He has published 30 articles in *QST* and won seven Cover Plaque Awards. John has conducted over 100 free Zoom presentations for radio clubs and presented at three QSO Today Virtual Ham Expos. A free list of presentation topics is on his website, **w6nbc.com**. He's active on HF, VHF, UHF SSB, FM, digital modes, and ham satellites, and he is a steam railroading, pipe organ, and sushi enthusiast. He's married to Rebecca Humes, KF6OEB. John can be reached at **jportune@aol.com** and 805-406-6176.

# The 2-Meter Slot-Cube Antenna

*Get high performance from a small package with this low-profile antenna made from copper pipe.*

John Portune, W6NBC

Few homebrew antennas have been more popular than the 2-meter J-pole made from ordinary copper water pipe. However, the J-pole's highly visible profile may be objectionable to some neighbors. If that's your situation, this may be the antenna for you. While still using copper pipe, you can build a less visible antenna with better performance than a J-pole.

## The Slot Cube

The diminutive 10 × 10 × 7 inch 2-meter base-station antenna shown in Figure 1 has the omnidirectional gain of a J-pole, but it is more efficient. No, it is not a halo — it is a folded skeleton-slot antenna. Halos are normally horizontally polarized. This antenna is vertically polarized by feeding the top and bottom loops in parallel (see **Figure 1**). Incidentally, this is a great antenna for 2-meter repeaters, which require vertical polarization.

The copper pipe loop portion of the skeleton-slot antenna is the equivalent of the solid plane in a classical slot antenna with the vertical space between the loops forming the slot. The whole antenna is folded vertically into a cube to make it omni-directional. For high efficiency and good bandwidth (see **Figure 2**), the conductor is comprised of common ½-inch copper pipe and copper fittings, which are available at most hardware stores and

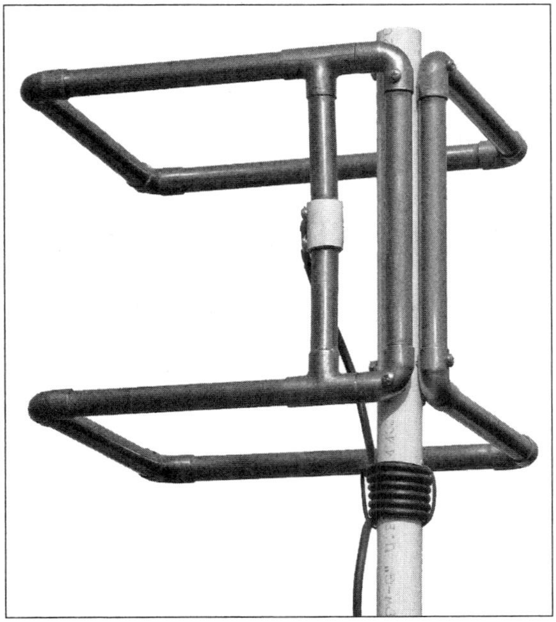

**Figure 1** — The 2-meter slot-cube antenna constructed from ½-inch copper pipe and plumbing fittings. The two horizontal loops act as planes with the gap between the two vertical connecting elements forming the slot. The feed point is at the center of the third vertical element connecting the two loops. Six turns of coax feed line around the mast form a choke balun.

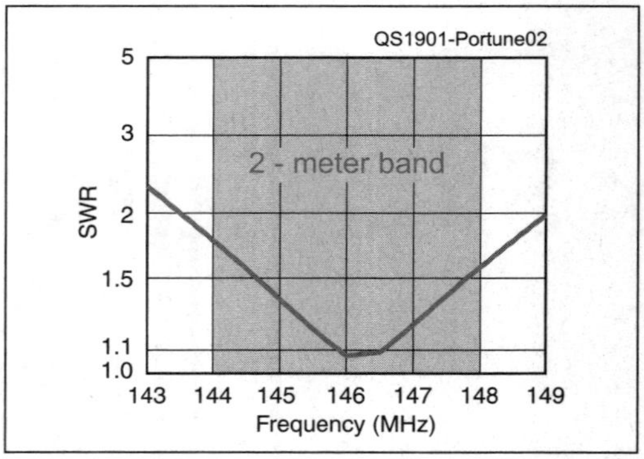


**Figure 2 — SWR plot of the slot-cube antenna.**

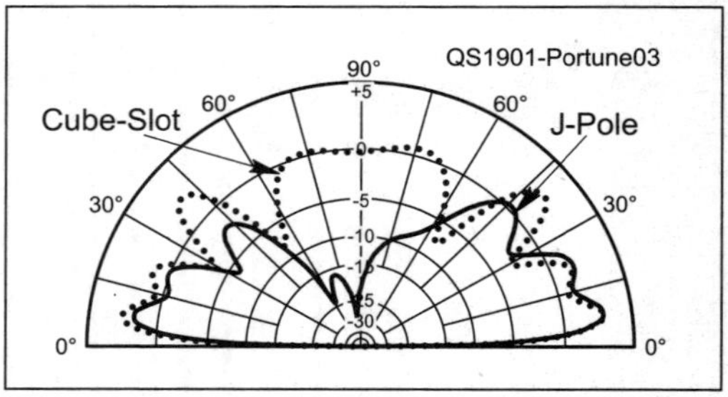


**Figure 3 — Vertical radiation plot of the slot-cube antenna (dotted) compared with a J-pole antenna (solid).**

home improvement centers. The average gain feature of EZNEC antenna modeling software shows a higher overall efficiency than a copper J-pole (see **Figure 3**).

## Construction

Rather than working from a dimensioned drawing, a better way to construct the antenna is to simply cut the copper pipe sections to length, as shown in the "Materials List" sidebar. The pipe sections naturally fit together to form the cube. Use Figure 1 as an assembly guide. It's a good idea to dry fit and mark all the pieces before soldering.

Begin assembly by cleaning the copper parts with steel wool. Then, separately solder together only the top and bottom loops on a flat surface using acid flux solder. Do not add elbows to the short ends yet. Next, assemble and solder the vertical sections. The already-soldered loops will hold the antenna in position for final soldering.

### Materials List

(4) 8½-inch-long horizontal pipe sections
(4) 8-inch-long shorter horizontal pipe sections
(2) 7-inch-long vertical pipe sections
(2) 3¼-inch-long feed point halves
(10) pipe elbows
(1) pipe T
(1) CPVC coupling

## Feed Point

Originally, I intended the feed point section to remain unsoldered and to slide for SWR adjustment. This required tedious filing out of the stops in the copper Ts to make this possible. Subsequently, I found that I could permanently solder the feed point section at the ends of the short loop sections.

There is a gap in the copper pipe in the feed point section inside the plastic coupling. Note that the coupling is CPVC, not common PVC. CPVC fittings are made to mate with copper pipe and will tolerate soldering temperatures. Connection to the antenna is made by drilling and tapping the CPVC coupling and copper pipe ends for 6-32 or 8-32 threads. Threaded holes are required for a positive connection.

Make up a short pigtail of RG-58 or Mini-8 coax with a connector on the end. Allow enough coax for a six-turn coax choke balun (see Figure 1). Secure it with a UV-stabilized tie-wrap through holes in the PVC mast. Weatherproof the coax ends with heat-shrink tubing and silicone sealant. For the remainder of the run to your rig, use larger-diameter, lower-loss coax, such as RG-8, RG-213, or LMR-400.

## Mounting and Disguise

If you pole-mount the antenna, drill four $^{11}/_{32}$-inch holes through the elbows and ¾-inch PVC mast section for 8-32 × 2 inch screws. Offset the holes to allow the mounting screws to pass. Alternately, the antenna may be set directly on your roof, keeping it level. Covering it with a small plastic dishpan or other plastic box will further obfuscate its function.

Previously, I published an article on a small compact transmitting loop antenna disguised as a working weathervane.[1] I consider the slot-cube antenna an improvement — it has superior performance and is far easier to build, while at the same time being suitable for the same method of disguise.

**Notes**

[1] J. Portune, W6NBC, and F. Adams, WD6ACJ, "A Weather Vane Antenna for 2 Meters," *QST*, Aug. 2005, pp. 35 – 38. Also available in the ARRL publication *Small Antennas for Small Spaces* by S. Ford, WB8IMY, or on the author's website, **w6nbc.com**.

Photos by the author.

John Portune, W6NBC, is an ARRL member and frequent contributor to *QST*. He has been licensed for 52 years and has held an Amateur Extra-class license since 1972. John has a BS in physics and also holds FCC Commercial General Radiotelephone Operator and FCC Radiotelegraph licenses. He retired as a broadcast television engineer and technical instructor at KNBC in Burbank and then from Sony Electronics in San Jose, California. You can reach John via email at **jportune@aol.com** or through his website at **w6nbc.com**.

# Slot-Cube Antenna for 6 Meters

***A scaled-up and improved 6-meter version of the popular 2-meter copper-pipe slot-cube antenna.***

John Portune, W6NBC

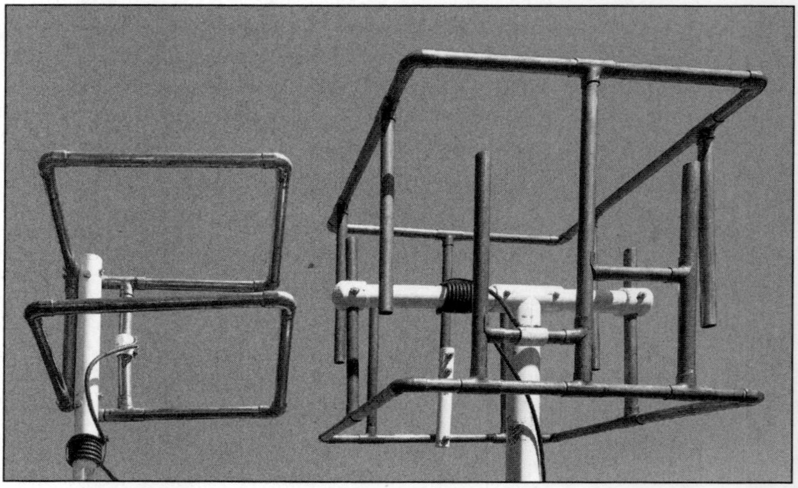

**Figure 1 — 2-meter and 6-meter slot-cube antennas.**

In the January 2019 issue of *QST*, I introduced the slot-cube, a remarkably small 2-meter base station antenna, with similar gain to a J-pole, and, like the popular home-brew version, soldered together from ½-inch copper water pipe. In response, I've received a flood of emails and photos from around the world, and many have wanted to know if the design can be scaled up to 6 meters. **Figure 1** shows my 16-inch 6-meter slot-cube antenna, only 60% bigger than its popular 2-meter cousin.

## How Slot Cubes Work

Slot cubes are folded *skeleton slot* antennas bent into a cube to reduce size. The Japanese "Hentenna," found widely on the internet, is a straight version of the skeleton slot. To adapt the 2-meter design to 6 meters, one needs to reduce the size even more. Simply scaling up the 2-meter slot cube in the same copper pipe makes for a heavy and large antenna on 6 meters. Therefore, I folded the slot one more time, adding open-ended vertical elements with caps (for rain: not yet installed in the figures) to stagger the slot. The slot now wanders up and down as it folds around the cube. **Figure 2** shows the evolution and size reduction, compared to its larger parents of a straight skeleton slot and a single-folded slot-cube, all on the same frequency.

## Performance

**Figure 3** compares the free space elevation radiation patterns of a normal 14-foot-tall, 6-meter J-pole (dotted) made from ½-inch copper pipe and the 16-inch double-folded slot cube (solid). Both are almost perfectly omnidirectional in azimuth.

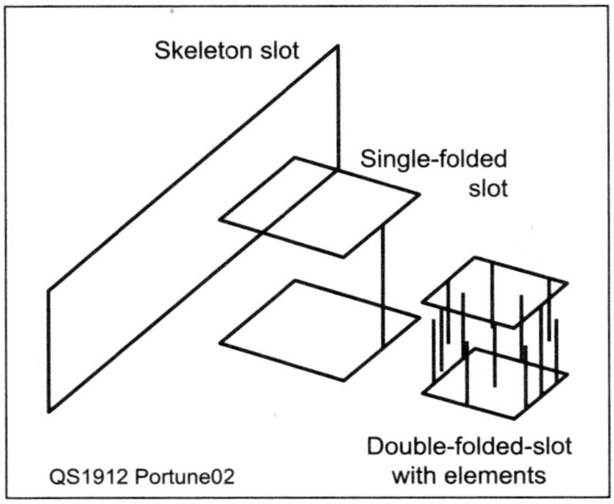


**Figure 2 — Evolution of 6-meter skeleton slot with relative sizes to scale, all on 6 meters. From left to right: unfolded, folded, and folded and staggered.**

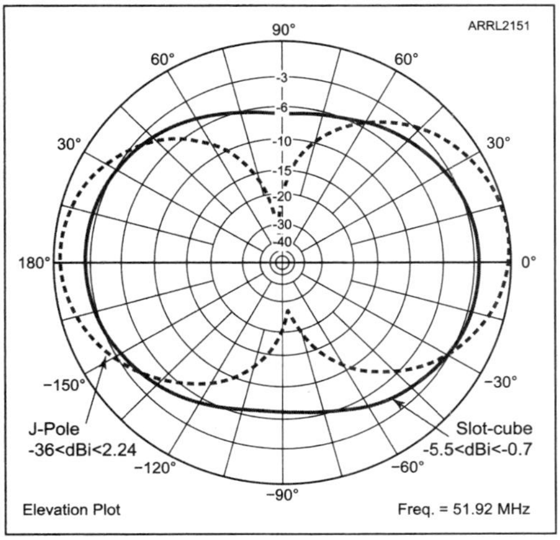


**Figure 3 — *4NEC2* copper pipe 6-meter J-pole (dotted) and slot-cube (solid) free space elevation radiation patterns.**

As seen, the slot cube has a nearly isotropic pattern, meaning it puts a signal in all directions. This requires that the gain towards the horizon is about ½ to ⅓ of an S-unit less than the J-pole, which has nulls vertically like any vertical dipole.

Efficiencies are above 95% for both antennas due to the large-diameter copper conductors. Like all size-reduced antennas, the slot cube is narrow in SWR bandwidth. The J-pole bandwidth is roughly 3 MHz at 2:1 SWR, whereas the slot cube is roughly 300 kHz. This means the slot cube must be carefully tuned to your operating frequency. Provision for that is provided in this improved design.

## Construction

The cube has two distinct halves (black and gray, see **Figure 4**) connected by a single vertical pipe (dotted). Solder together each antenna half separately. Four common bricks laid on a flat concrete surface work well during soldering. Square up everything before beginning. Misadjustments are, however, easy to correct by reheating individual joints after the cube is fully assembled. At first, do not solder either end of the connecting section (dotted). You may wish to change its length during tuning. The antenna will function normally when only partially soldered, and final soldering can be done after tuning. The 7-inch black piece is the PVC brace.

## Improved Tuning

The tuning and SWR adjustments are easier here than on the 2-meter version. I added movable pipe sections at both ends of the slot (see **Figure 5**). I made the movable pipe ends by cutting a copper T in half longitudinally through the center of the side arm. I further cut off the corners at a 45-degree angle. Then, I filed out any internal material to allow the curved surface to slide freely along the tubing. The short cross-pipe section on the left tunes the frequency. If the overall dimensions are correct, it should be possible to tune over the 50 – 54 MHz range.

The section with the CPVC coupling adjusts the SWR (the match). Adjusting the tune frequency affects the match only slightly. The

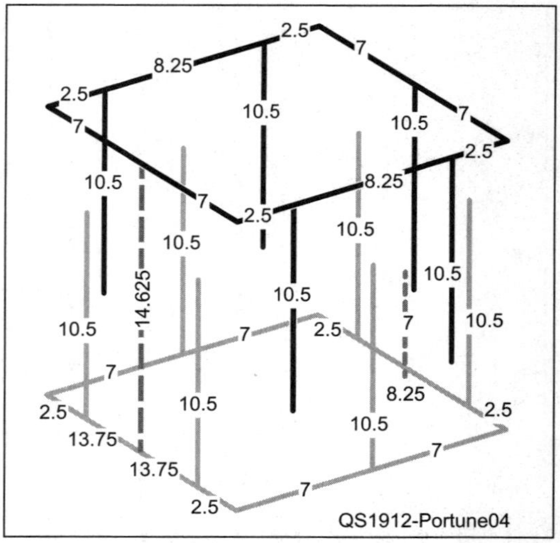


**Figure 4 — Cut lengths (inches) and locations of copper pipe sections.**

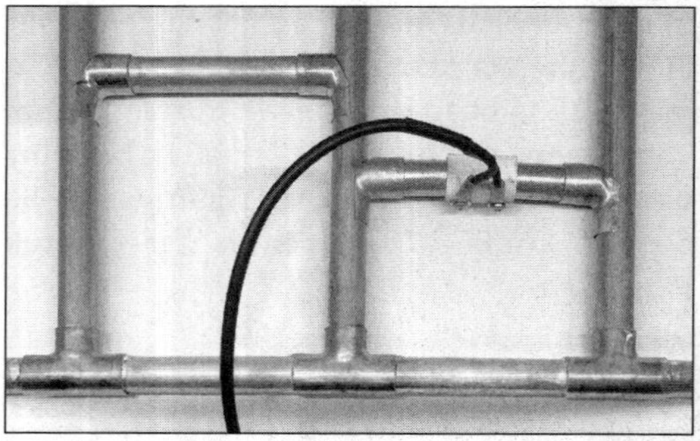

**Figure 5 — Movable bars, tune (left), and match (right). Also, note the feed point.**

feed coax attaches on this section by crimped-on ring terminals and 6-32 screws into tapped holes in the pipe ends. Weatherproof the separated coax wires with heatshrink tubing. There is roughly a ⅛-inch gap between the pipe ends inside the CPVC coupling. After tune and match adjustments at the working location of the antenna, the movable sections may be screwed, hose-clamped, or soldered in place.

## Materials List

The dimensions are cut lengths of the pipe sections, not overall antenna dimensions. They do not include the length added by the elbows and Ts. Assemble the antenna with these pipe lengths and the tuning will be correct.

(8) 2¼-inch lengths of ½-inch rigid copper water pipe

(2) 3¾-inch lengths of ½-inch rigid copper water pipe

(8) 7-inch lengths of ½-inch rigid copper water pipe

(3) 8¼-inch lengths of ½-inch rigid copper water pipe

(11) 10½-inch lengths of ½-inch rigid copper water pipe

(1) 14.625-inch lengths of ½-inch rigid copper water pipe

All of the above can be cut from two 10-foot lengths of copper pipe.

(8) ½-inch short-radius (standard type) copper pipe elbows

(15) ½-inch short-radius (standard type) copper pipe Ts

You must use standard radius parts (not extended radius), for correct dimensions.

(8) ½-inch copper pipe caps for open-ended pipe sections

(1) CPVC pipe coupling (not a PVC coupling, which won't fit the tubes)

(2) crimp terminals for #6 stud, to be installed on coax

(8) 10-24 × 2 inch stainless-steel screws and nuts

(2) 6-32 × ⅜ inch stainless-steel screws

(2) 6-32 stainless-steel flat washers

(1) RG-8X or RG-58 for feed and balun

(1) 5½ inch × 1 inch PVC pipe

(1) 9 inch × 1 inch PVC pipe

(1) 1-inch PVC T

(2) 1-inch PVC caps for mounting (see "Mounting" section)

(1) 7 inch × ½ inch PVC pipe, for brace

Figure 4 is the physical layout of the antenna, including the locations of the copper-pipe sections listed.

## Soldering Tips

The antenna may be assembled by following standard plumbing methods with sweating flux. Excess solder will not affect performance but may be unsightly. For a cleaner result, the following alternate process may be used.

1 Degrease pipe sections and fittings with acetone.

2 Carefully remove copper patina with swimming pool hydrochloric acid and rinse thoroughly.

3 Apply a coating of rosin flux thinned with acetone. Let the flux dry.

4 Assemble and add a minimum amount of solder. Unlike with plumbing work, the joints don't need to be watertight.

5 After you have tuned the cube in its working location, you may just temporarily remove any non-copper parts, reheat the joints, straighten things up, and add more solder.

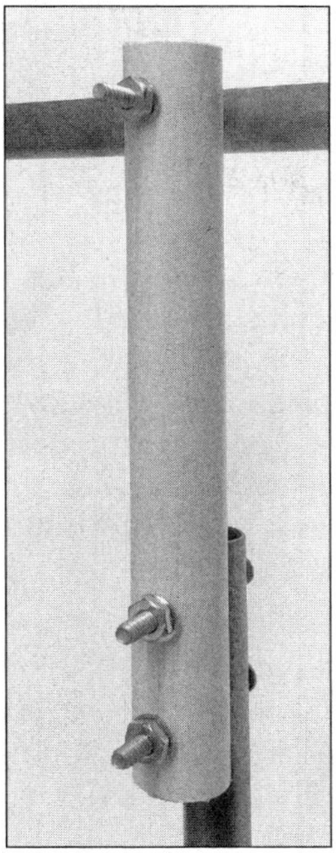

Figure 7 — Mounting brace location.

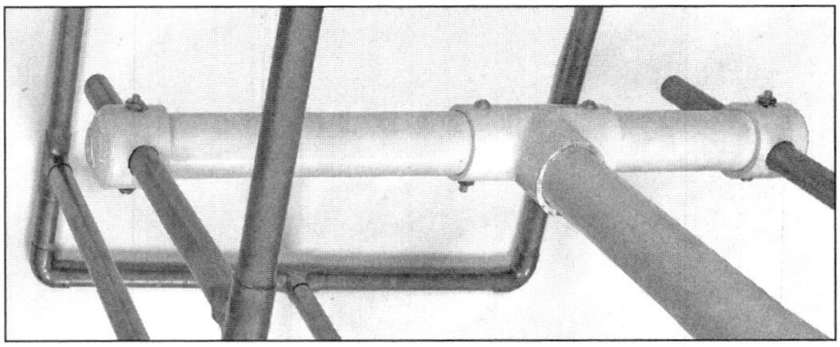

Figure 6 — Mounting assembly. Note ⅝-inch holes through the end caps and 10-24 x 2 inch screws.

## Mounting

One-inch PVC is suitable for mounting (see **Figure 6**). The ⅝-inch holes drilled through the end caps allow the mount ends to slip over two of the open-ended pipes. The mount is centrally located in the cube and secured by 10-24 × 2 inch stainless-steel screws through the end caps and pipe. An additional small seven-inch brace of ½-inch PVC pipe provides final stabilization for the cube, as seen in **Figure 7**. It is attached on the opposite face of the cube from the crossover pipe by three 10-24 × 2 inch screws.

The main mounting T is located slightly off center to allow the antenna to be mounted for either vertical or horizontal polarization. As oriented in Figures 1 and 4, the polarization is vertical, which is preferred for FM work, and will work fine for DX work (sporadic E) on all modes via the ionosphere. For local SSB contacts, horizontal polarization is preferred. Further details on patterns and mounting for horizontal polarization can be found on the "*QST* in Depth" web page (**arrl.org/qst-in-depth**).

For a mast or mounting nipple, use Schedule 40 PVC or aluminum pipe or a steel TV mast inserted into the mounting T secured with a 10-24 × 2 inch screw through the mast.

Lastly, a 1:1 current choke balun is required near the antenna. A handful of Mix 61 ferrite toroid cores slipped over the coax may be used, or nine turns of RG-8X or RG-58 around the PVC cross-arm or mast secured with cable ties through 3⁄16-inch holes. The balun prevents detuning and distortion of the radiation pattern by common-mode currents in the coax and should be present when doing adjustments, as it may affect the measured SWR.

---

Photos by the author.

John Portune, W6NBC, is an ARRL member and frequent contributor to *QST*. He has been licensed for 54 years and has held an Amateur Extra-class license since 1972. John has a BS in physics and also holds FCC Commercial General Radiotelephone Operator and FCC Radiotelegraph licenses. He retired as a broadcast television engineer and technical instructor at KNBC in Burbank and then from Sony Electronics in San Jose, California. He will be pleased to correspond via email at **jportune@aol.com**, or at **w6nbc.com**.

CHAPTER 3

# Disguised and Novelty Antennas

# A 160 or 80 Meter Downspout Vertical

*It doesn't get much stealthier than this — make your downspout do double duty.*

Dave Holdeman, N9XU

Figure 1 — View of the toilet bowl mounting flange and small joiner as purchased. Note that the PVC toilet bowl mounting flange has a stainless steel ring around the outside with six holes.

## The Past

While cleaning out the basement, I ran across some old notes and logs from 1958 and 1959, my Novice days. What caught my attention was the antenna I used back then. I believe it came from an old article from *Popular Electronics* magazine. It consisted of a 12 foot piece of 2 × 2 inch lumber with holes drilled through it every 1.5 inches. Wire sized #14 AWG was then run in and out of the holes from one end of the 2 × 2 to the other. This made the electrical length of the antenna longer.

On one end I mounted a copper toilet bowl float for a capacitance hat and on the other end I attached an old military surplus inductor that I equipped with taps. I mounted the whole affair vertically on the side of the house with the capacitance hat up and proceeded to work 46 states, mostly on 40 meters.

## The Present

The fine article by Greg Crossman, WEØD, in the August 2010 issue of *QST* stirred my thinking and caused me to consider an antenna for 160 meters, a band that I had never worked.[1] While I didn't need a stealth antenna as Greg did, I liked his idea of using an electrical junction box to terminate his antenna radials. Greg's article also made reference to the excellent article by Rudy Severns, N6LF, on ground radials for HF vertical antennas.[2] That was all it took to get me started.

## Overview

I wanted to construct the antenna out of lightweight material that could be built and put into position by one person. I also liked the in and out feature of the old 2 × 2 antenna wire and the simplicity of the toilet bowl float as a capacitance hat. Coincidentally, my son and I had just finished a sun porch addition to the house, which we equipped with 2 × 3 inch plastic downspouts. I decided to build the antenna using two 10 foot sections of that material.

The top section would contain the capacitance hat and the bottom section would contain a base loading coil and electrical junction box to terminate the ground radials in a manner similar to that described in Greg Crossman's article. I was unsure how I was going to build and mount the base coil, but a trip to the hardware store solved that and other problems.

## The Hardware Store

I am always amazed at what one can find at a hardware store that is adaptable to ham radio. I first went to the electrical section and picked up a 500 foot roll of #14 AWG stranded wire, an electrical box, two 8 foot ground rods and four ground clamps. I then headed for the plumbing section and happened to spot a toilet bowl mounting flange made of heavy PVC. This could be used to mount the capacitance hat and six hat radials made out of 1⁄16 inch brass rod that I found at a hobby shop. I looked for a copper toilet bowl float, but only found plastic ones until I went to a store that had been around for a long time and had old stock.

I found the downspouts and two types of joiners that are normally used to connect two downspout sections end to end. One type of joiner was socketed and that was the one I used to join the two sections. I put the two sections and the joiner together and picked up the whole assembly by the middle. While the ends drooped, it appeared that by strategic use of self-tapping screws, the joint would hold, especially if mounted vertically. I planned to use the other, shorter, joiner to mount the capacitance hat assembly.

I also needed something on which to wind my base coil. In the same area I noticed some round flexible downspout about 4¼ inches in diameter that could be stretched or compressed. This was ribbed and would hold coil turns firmly and neatly. I added two of those to my cart and headed for the checkout lane.

The capacitance hat consisted of the PVC toilet bowl mounting flange, the toilet bowl float, six brass rods for top radials, and a short downspout joiner. See **Figure 1** for a picture of the toilet bowl mounting flange and small joiner as purchased. You will note that the PVC toilet bowl mounting flange has a stainless steel ring around the outside with six holes. I covered the top of the PVC area with masking tape and drew lines from three of the six holes to their opposite number. Where the lines crossed is the exact center of the toilet bowl mounting flange. I drilled a hole at this point large enough to accommodate a ¼-20 machine screw. This will secure the copper toilet bowl float and top radials when assembled. I removed the masking tape. I then created six top radials 18 inches long from 1⁄16 inch brass rod by soldering a terminal lug with a ¼ inch hole to one end of each rod.

Adjacent to the holes in the stainless steel ring on the toilet bowl mounting flange, there is a circular ridge around the circumference of the PVC. I drilled six equally spaced 1⁄16 inch holes through the ridge right at the spot where the holes are in the stainless steel ring. These holes would be used to anchor the six brass rod radials. I then removed the stainless steel ring from the PVC as it was no longer needed. I temporarily bolted the toilet bowl float, radials, and PVC flange together so it looked like the assembly shown in **Figure 2**. I still had to attach the capacitance hat to the top of the rigid downspout. I disassembled

Figure 2 — Assembled PVC flange with attached capacitance hat.

Figure 3 — Bottom view of the capacitance hat assembly.

Figure 4 — The top section of the assembled antenna with the capacitance hat attached.

the capacitance hat so as not to damage the radials and turned over the toilet bowl flange. I drilled two holes in the side to attach the small downspout joiner. I used two long #6-32 machine screws, nuts, washers, and plastic spacers. Refer to **Figure 3** for a bottom view of the capacitance hat assembly. Reassemble the unit, carefully drill a couple of drain holes in the top, sit back and admire your work and lay the capacitance hat assembly aside.

I am sure that the antenna would work well without the toilet bowl float, but the top hat radials should remain. Since the toilet bowl float is a scarce item anyway, it could be eliminated. It could be replaced by an acorn nut and washer that would still hold the capacitance hat radials in place.

## The Top Section

I moved my wife's car out of the garage and used that area to continue building the antenna. I took one 10 foot section of downspout and laid it across a pair of sawhorses with the wider side up. I ran a strip of masking tape down the length of the downspout and marked a spot every 1½ inches. I turned the downspout section over and did the same to the opposite side. I drilled holes in both sides large enough to accommodate the #14 AWG wire.

I removed the masking tape and threaded the wire in and out of the holes starting from one end and continued until I reached the other. I left about a foot of extra wire at each end. I determined which end of the downspout would be at the top and connected the wire to the top hat assembly using a ring terminal lug with a ¼ inch hole. I used PVC cement to secure the top hat assembly to the downspout, pushing the excess wire down into the downspout. The top section with the capacitance hat attached is shown in **Figure 4**.

## The Base Coil Form and Bottom Section

The base coil is wound on a form made of two pieces of ribbed flexible downspout joined together. One piece of flexible downspout (as purchased) is shown in **Figure 5**. Each flexible piece has an end that was made to fit over the 2 × 3 inch rigid downspout. The other end of each flexible piece didn't match anything particularly well, so I cut those ends off both samples using a sharp utility knife. I also cut off the 2 × 3 inch end of one of the flexible pieces.

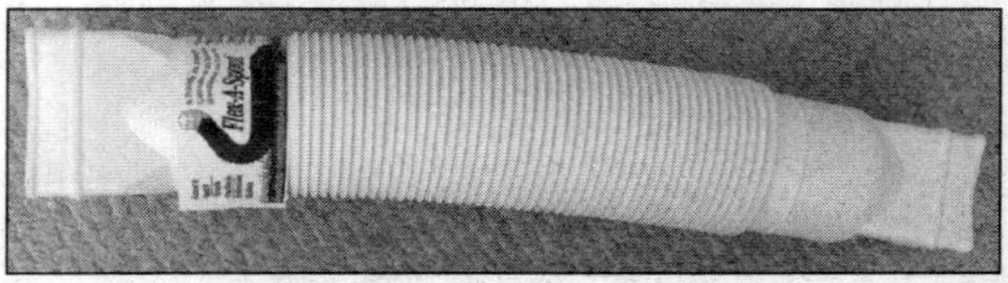

Figure 5 — The flexible downspout section used as a form for the loading coil.

Figure 6 — The completed base coil. Note the slits that allow the wire to transition between grooves.

Figure 7 — The junction box assembly along with two of the 4 foot ground rods.

I laid the remaining 10 foot section of rigid downspout on my sawhorses and slid the flexible piece of downspout that still had the 2 × 3 inch end on it over the rigid piece. This would become the top half of the base coil. I slid the remaining flexible downspout onto the rigid section and joined the round ends together to complete the base coil form. I found that one end slipped into the other snugly; I secured them with self tapping screws. I used white self-tapping gutter screws that matched nicely. You will find that the whole coil form assembly will slide up and down on the rigid downspout at this time. It will be affixed to the rigid downspout later.

The entire antenna assembly would be set in a 2 foot deep hole. I mounted an electrical junction box about 28 inches above the lower end of the bottom section. The junction box contains an SO-239 UHF coaxial jack on the bottom and a ceramic feed through insulator on the top. The junction box would also be used to terminate my radial field and ground rods later on.

Once the junction box was mounted, I secured the base coil form to the rigid downspout just above the ceramic insulator on the top of the junction box. I used self-tapping gutter screws at the top and longer self-tapping screws and plastic spacers at the bottom. Once the base coil form was attached, I drilled more holes 1.5 inches apart in the remaining part of the rigid section above the base coil form to pass the additional wire needed to connect to the upper section.

## Winding the Base Coil and Final Assembly

The ribs on the base coil form are not spirally wound, but are concentric. I cut a notch in each rib to allow the #14 AWG wire to pass from one groove to the next. I staggered the notches so the coil windings would not bunch up in one spot on the form.

Winding the wire on the coil form was not difficult since the ribs held the wire nicely as I wound. Initially, I didn't have a clue as to the antenna's resonant frequency, so I wound as many turns on the form as possible, figuring I could take excess turns off later as needed. The completed base coil is shown in **Figure 6** and the junction box assembly is shown

Figure 8 — A portion of the lower section and brace.

in the ground with stones to make the bottom much more rigid and I breathed easier. I later added a brace made out of PVC pipe and an aluminum bracket between the house and the lower section. A portion of the lower section and brace is shown in **Figure 8**.

in **Figure 7**. Figure 7 also shows two of the 4 foot ground rods mentioned later on.

## Erecting the Antenna

With safety in mind, I moved everything out of the garage into the backyard and supported the two sections on the sawhorses and a couple of lawn chairs, with the bottom end of the first section close to the hole in the ground that was to receive the base of the antenna. I connected the two sections together through the longer joiner and secured the joint with six self-tapping screws. I connected the wires together from the two sections with wire nuts. I put my extension ladder in place against the house rain gutter, with the base of the ladder close to the hole in the ground, and made ready two large C clamps to temporarily hold the antenna upright.

I then horsed the antenna upright and the base slipped into the hole with a satisfying thunk. Using the C clamps, I temporarily clamped the antenna to the extension ladder and the upper section to the house rain gutter and stepped back to catch my breath. I loosened the C clamps alternately as needed and shifted the ladder in small increments to make the antenna as vertical as possible.

I bolted the antenna to the rain gutter using two long #6-32 screws, ¼ inch plastic spacers, and airplane nuts. I filled the antenna hole

## RF Safety and RFI

Any antenna that is built up against the house could radiate into an occupied area or subject people or animals outdoors to RF. It is thus particularly important to perform an RF safety assessment, as required by the FCC. I am running no more than 100 W of power output with low duty cycle CW and SSB operation, so RF exposure issues are not an issue at my station.

In addition, because of the proximity to the house and its systems, RF can couple into plumbing, power and communication wires, or video cable lines. If the location of these can be ascertained, it is best to position the antenna as far from them as possible. When I had my house built, I had the contractor put the telephone and electrical wires into separate conduit. To date I have not experienced any RF problems.

## The Ground System

I buried five 160 meter radials with the first two running along the side of the house in opposite directions and around toward the front yard. The other three are more straight and conventional. Each 160 meter radial is 86 feet long and is buried about 6 inches in the soil. I made the radials of the same #14 AWG insulated stranded wire as is used in the antenna. This is shorter than the standard

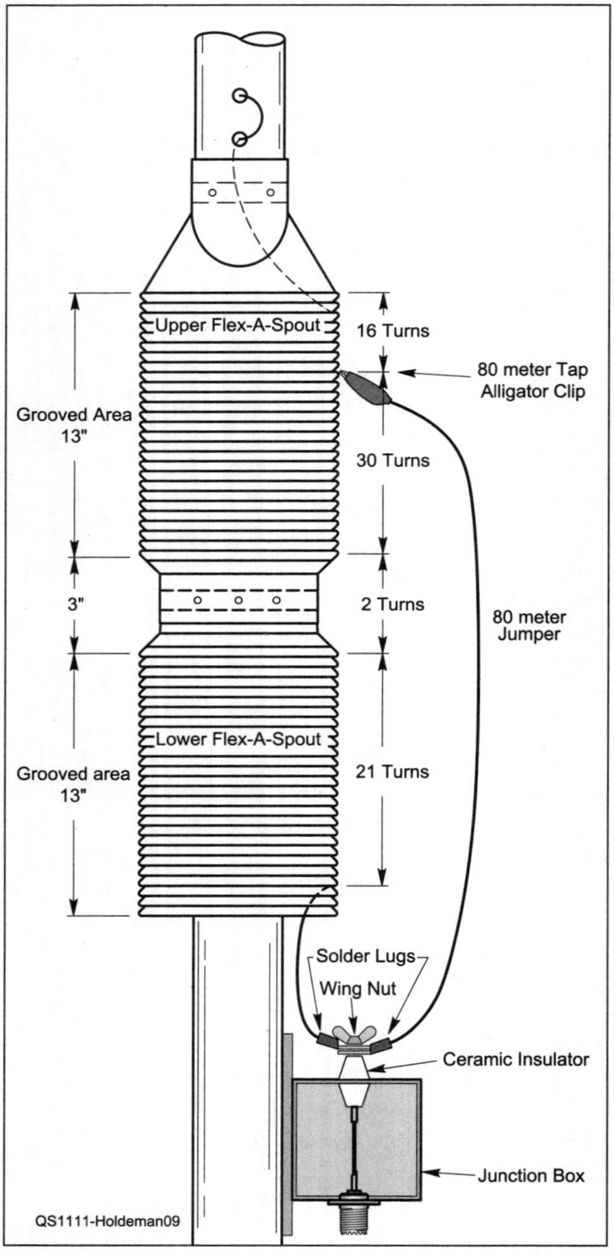


**Figure 9 — The completed base coil showing the number of turns of wire required for both 160 and 80 meters.**

formula, but was the result of reading Rudy Severns' experiments on ground radials. I cut the two 8 foot ground rods in half and created four shorter ones out of the original two. (I have had sad experiences driving 8 foot ground rods down through Illinois clay.) I connected the radials and ground rods to the junction box and connected up my MFJ antenna analyzer, wondering where I would see a dip.

## Resonance

I saw the start of a dip right at 1.6 MHz, the extreme end of the analyzer's range. I removed turns from the bottom end of the base coil until the indicator dipped at 1.845 MHz. Although I had a very pronounced dip, the analyzer showed only a 2.5:1 SWR with a 27 Ω impedance at that point. I took the analyzer into the shack and ran it through my MFJ 949D antenna tuner. I was able to tune to a 1.1 SWR from 1.805 to 1.850 MHz.

## The 80 Meter Option

I decided to see how my new addition would perform as an 80 meter antenna. I unwound turns from the bottom of the base coil until the analyzer showed resonance at the middle of the 80 meter band. The analyzer showed a very nice 1.1 SWR with an antenna impedance of 50 Ω at 3.825 MHz. With my antenna tuner, I was able to get a 1.1 SWR across the entire band. I marked that portion of the base coil with a wire nut and buried five more radials cut for 80 meters in between the 160 meter radials. These radials are 43 feet long and are also made of #14 AWG wire. I rewound the bottom portion of the base coil with new wire to get back to my 160 meter resonance point. I modified the top of my feed-through insulator on the electrical junction box with a wing nut so I could attach a jumper to connect to the 80 meter tap on the base coil as desired. **Figure 9** depicts the completed base coil and the number of turns of wire required at my station for both 160 and 80 meters.

## On the Air

I first tried out the antenna in the October Illinois QSO Party. I had worked several stations on 40 meters in the afternoon with my 40 meter G5RV, so after supper I dashed outside to set the 80 meter tap on the vertical. I worked eight stations on 80 meter CW in a short period of time. I was just getting warmed up, but the band suddenly started to go quiet and I realized that the contest was over. I didn't get a chance to work anyone on 160 meters that day, but I have worked several stations at random since the contest and I am happy with the results.

## Possible Variations

If one were to construct the antenna with white wire, the wire would be barely noticeable. Elimination of the copper toilet bowl float, along with the wire change, would make it very nearly a stealth antenna.

If you are concerned about weather effects on the junction box, it is possible to obtain one made of ⅛ inch gray plastic with an airtight lid designed for marine applications. It also has four mounting tabs. One could mount an aluminum plate inside and terminate the radials through the side of the box with screws. One would also have to ground the shell of the SO-239 jack to this plate.

---

**Notes**

[1]G. Crossman, WEØD, "Earning 160 Meter WAS in 117 Days," *QST*, Aug 2010, pp 44-45.

[2]R. Severns, N6LF, "An Experimental Look at Ground Systems for HF Verticals," *QST*, Mar 2010, pp 30-33.

---

Dave Holdeman, N9XU, was born and raised in Muskegon, Michigan. After graduation from Muskegon High School, Dave joined the Army Signal Corps. He attended signal school in Ansbach, Germany and met his future wife, Christa, in Berlin while on temporary duty there. In the service he worked on BC610s, BC342s and other radio and cable carrier systems.

After his Army tour, Dave studied broadcast engineering received his FCC First Class Radiotelephone license and joined the staff of WONW-AM in Defiance, Ohio. After licensed engineers were no longer required at broadcast stations, Dave became a microwave communications craftsman at AT&T. He performed acceptance testing on new microwave radio installations and worked the microwave test bench.

In 1958 Dave passed his Novice exam and received the call KN8OIO. Later he passed his Technician and General class exams and became K8OIO. In 1963 he was promoted to AT&T Chicago Engineering and received the new ham call W9HJL. He now holds an Amateur Extra class license. You can reach Dave at 415 Barnaby Dr, Oswego, IL 60543 or at holdex@att.net.

# The Cookie Tin Antenna — Making it Work

*This familiar makeshift antenna may surprise you when it's configured properly.*

John Portune, W6NBC

My ham buddy, Ernie, W6ND, loves to "kluge" together projects from baling wire, wooden dowels, and PVC pipe. He and I are constantly at loggerheads over his makeshift designs (in fun, of course). Our latest dustup was over a 2 meter portable antenna he made from a 7½ inch cookie tin (see **Figure 1**). When I first saw it, I delivered the classic ham squelch: "The ground plane is too small!" But after seriously looking at cookie tin portables with *EZNEC* antenna modeling software and doing some live field tests, I discovered that an antenna that looks like a mechanical kluge can still be a decent electrical performer. You just need to know how to configure it.

## An *EZNEC* Model

Thinking that I could defeat Ernie with technical bluster, I organized my frontal attack around an *EZNEC* wire frame model of his antenna (see **Figure 2**). Actually, to gather the comprehensive data here, I made seven models: for 2½, 5, 7½, 10, 20, 40, and 80 inch diameter cans, all 2 inches thick. Some were theoretical, as you can imagine. I modeled them all at 146 MHz, 2½ feet (picnic table height) over average soil, and for a wave angle of 15°. I added an eighth model as a reference: a half wavelength vertical with no cookie tin.

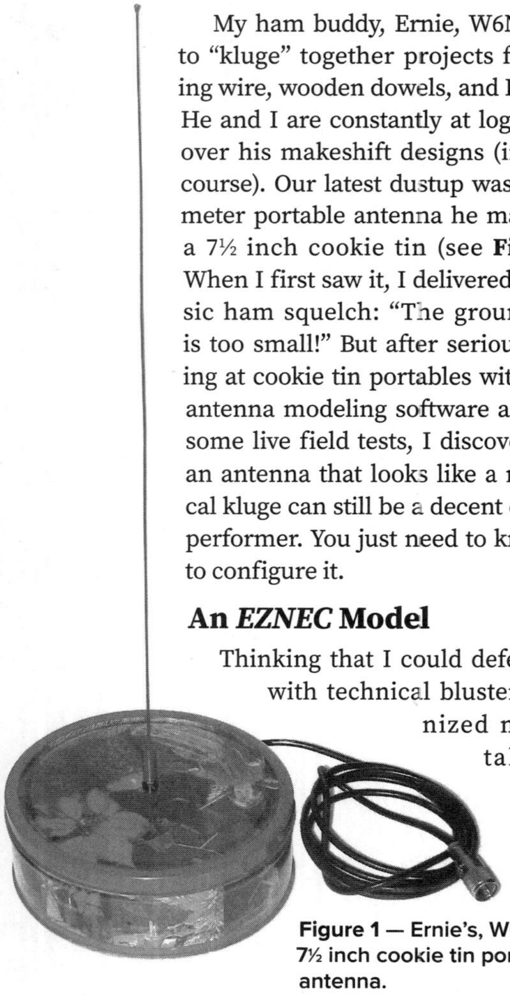

**Figure 1 — Ernie's, W6ND, 7½ inch cookie tin portable antenna.**

I thought I was ready to show Ernie that his 7½ inch cookie tin antenna wouldn't work well, but the data shown in **Table 1** and **Figures 3** and **4** surprised me. All three figures use the same *EZNEC* modeling data. Note the variation in the optimum whip length determined for each tin.

The data made it clear to me that an efficient cookie tin portable *must have a longer whip* if the tin is smaller. It's the same if you put the whip on any surface smaller than half a wavelength in extent (40 inches). Don't just plunk down a 19 inch antenna and expect optimum performance. RVers with smooth-sided motor homes, pay close attention here. You can actually use a fairly small metal plate underneath the whip on top of your rig and still get good performance. You must, however, increase the whip length.

Few hams, however, seem to grasp this principle. Most just assume that 19 inches is

**Table 1**

**EZNEC Predicted Performance versus Tin Diameter**

| *Tin Diameter (inches)* | *Antenna Length (inches)* | *Gain (dBi)* |
|---|---|---|
| 0 | 38.5 | 1.66 |
| 2.5 | 28.5 | 1.61 |
| 5 | 25.5 | 1.51 |
| 7.5* | 23.4 | 1.27 |
| 10 | 22.5 | 1.02 |
| 20 | 20.5 | 0.89 |
| 40 | 19.1 | 0.64 |
| 80 | 19.0 | 0.14 |

*W6ND's antenna

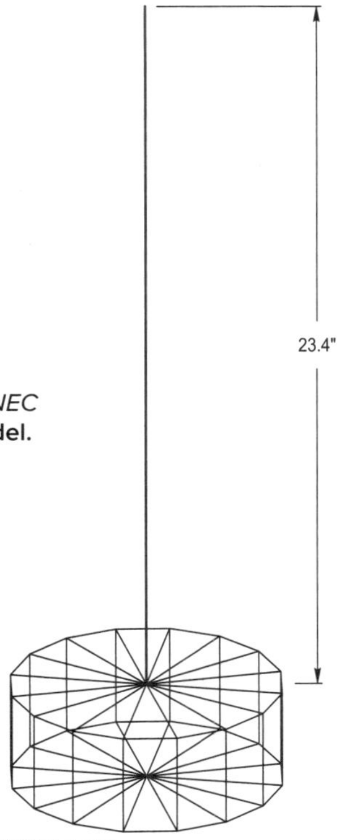


**Figure 2** — *EZNEC* wire frame model.

the only length a 2 meter whip should ever be, no matter where it is mounted. But making it the right length is the key to success with a cookie tin portable antenna. Figure 3 shows that as the tin diameter decreases, the antenna length should increase for resonance. Suppose you were to shrink the tin down to nothing. How long would you make the whip — still 19 inches? No! You must lengthen it to ½ wave or 38 inches. This is no surprise; the quarter wavelength whip has become a full half wavelength dipole.

Figure 4 further shows the gain of a cookie tin antenna with the correct whip length. Interestingly, it actually improves slightly for smaller tins. This is because the radiating part of the antenna is longer. The difference though is minor, just over 1 dB. If you don't increase the whip length, however, the performance will degrade. I have even seen hams surprised that a 19 inch antenna on a wooden bookshelf or on a smooth-sided motor home with no metal plate or ground plane didn't work well.

## Why This Works

You may still be thinking, "I don't believe this; you can't make the ground plane so small." Not so! A tin less than 40 inches in diameter (half a wavelength) is actually no longer a ground. Plane rules don't apply for small tins. At that point the tin is a capacitive hat. It has become end loading for a shortened half wavelength dipole rather than a reflective surface to provide a virtual image of the antenna.

I implement a 2 meter cookie tin antenna like this: First I replace the 19 inch whip with something a lot longer. An adjustable whip antenna is ideal. Otherwise I just temporarily twist or solder some solid wire onto the top of the antenna. Then I begin trimming in small increments, using an antenna analyzer to let me know when I have achieved a low SWR. When the lowest SWR has been found, the assembly is *on frequency* and it will radiate well. I also use a choke balun while tuning. Without the balun, the feed coax will become part of the antenna. I usually just wind several feet of the coax into a small multi-turn bundle near

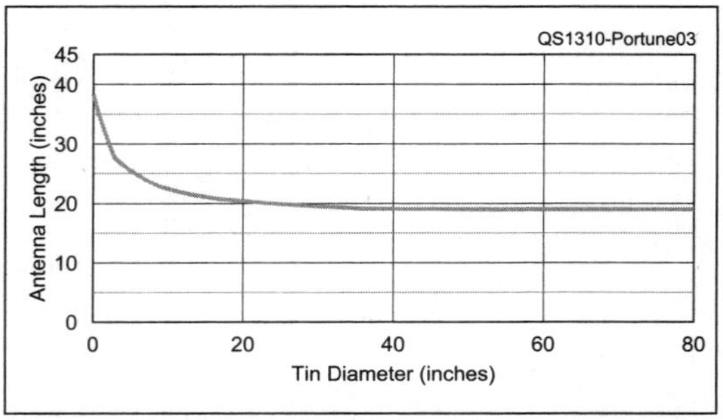


Figure 3 — Whip length to achieve resonance.

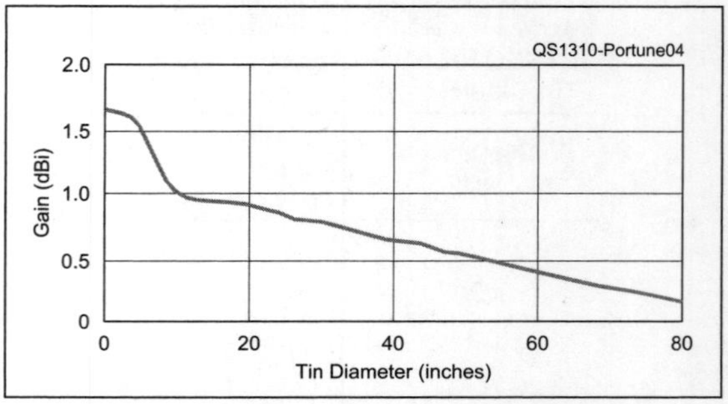


Figure 4 — Gain versus tin size at a 15° wave angle with whips of the correct length.

the tin or place ferrite beads on the coax (connectors permitting). I am also careful to keep the antenna away from metal objects, such as the supports for my workbench.

## A Coffee Can Portable

After recovering from Ernie's gentle scorn, I reasoned, "Shouldn't I now be able to make an effective portable from any can?" **Figure 5** shows what I "brewed up" from a coffee can and a magnetic mount.

Again, I first replaced the original 19 inch whip with 30 inches of #10 AWG solid copper wire, which is an adequate length for trimming. Then on a wooden TV tray 2½ feet off the ground, and with an MFJ-259B antenna analyzer, I shortened the whip in small increments until I reached the 2 meter band. At that point, the top was 25 inches above the coffee can — a full six inches longer than the original 19 inch antenna. I then replaced the copper wire with a stainless steel whip of the same length.

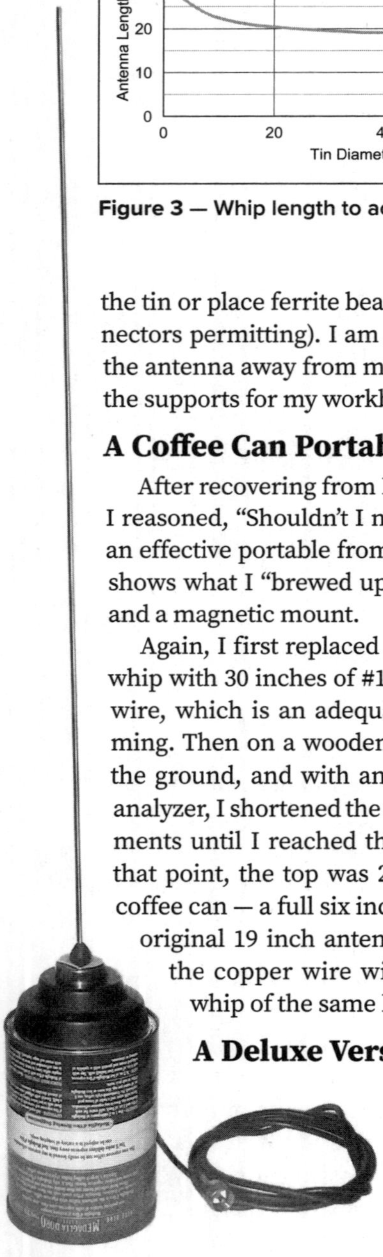

Figure 5 — Coffee can 2 meter portable antenna using a magnetic mount and a 25 inch whip.

## A Deluxe Version

For a fancier version I made a cookie tin portable from a 10 inch diameter can, a UHF plug and jack, and a small telescopic whip from a portable broadcast receiver (see **Figure 6**). This is my "use anywhere" antenna that I keep in my RV. I cut a slot in the side of the bottom of the tin so that I can roll up the coax and put it inside along with the whip when not in use.

If I'm using the antenna on a large metal surface I simply adjust the whip length to 19 inches. On a non-metallic surface it needs to be extended to roughly 22 inches. On other surfaces, the length is in between 19 and 22 inches and I use an SWR bridge or antenna analyzer to make the adjustment.

If you lack a cookie tin, just use a magnetic mount directly on top of a rig with a steel case. As with any metal mounting surface, the length of the whip is easily tuned to a proper length.

## Proof of Concept

To be sure that I wasn't just fooling myself with antenna modeling software, I decided to perform field tests. I have a 2 meter FM receiver calibrated in dBm. I connected it to a half wavelength vertical coaxial dipole at 20 feet. The test transmitter was a low-power transmitter with a 20 dB pad to ensure a 50 Ω source. It also kept the input signal below receiver saturation. The receiver was roughly 200 yards away from the transmitter, on the wooden TV tray 2½ feet above the ground.

◀**Figure 6** — Deluxe 10 inch cookie tin with telescopic whip in a UHF connector. Whip and coax are put inside the can for storage.

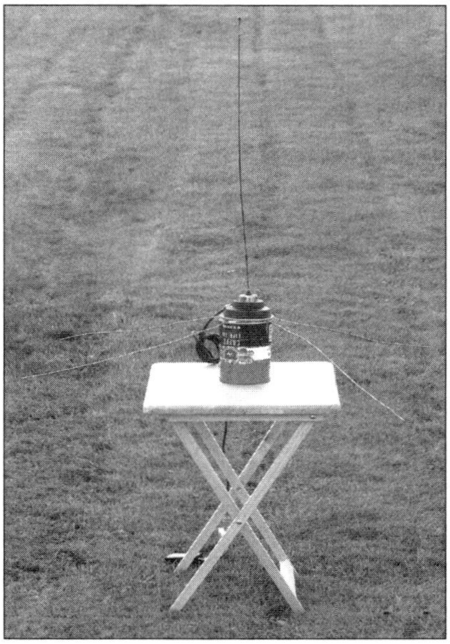

▶**Figure 7** — Coffee can antenna with 19 inch antenna and 4 drooping 21 inch radials.

I chose the coffee can magnetic-mount antenna for the tests. A cookie tin antenna would have been very similar. It was configured in three ways: (1) with only the original 19 inch antenna; (2) with the retuned 25 inch antenna; and (3) with the original 19 inch antenna, but this time with four added drooping radials of #10 AWG solid copper wire. The latter was an improvised ground plane antenna. It had essentially the same radiation pattern and gain as a vertical half wavelength dipole or a J-pole. I pruned the radials to bring the whole antenna to tune. The radials were roughly 21 inches long (see **Figure 7**).

**Table 2** compares the three. Both the ground plane antenna and the coffee can with the correct antenna length were within ½ dB. The coffee can with the original 19 inch antenna was 1½ dB weaker. Lengthening and retuning definitely restores the performance.

## What About Matching?

Matching was a surprise. The SWR actually changed little when I lengthened the whip on a smaller can. For example, the feed points in these simulations were dictated by the base mount. They were all at the bottom of the whip, just above the can. Even so, the SWR changed little. In fact, the smaller tins had better SWR. The reason, I believe, is as the whip gets longer, the feed point essentially moves away from the electrical center of the antenna. This raises the feed impedance, thereby compensating for the change of length.

Never again will I think of grandma's cookie tin as a ground plane unless it is larger than 40 inches in diameter. (I wonder how many cookies such a tin would hold!)

**Table 2**

**Field Test Results**

| *Antenna* | *Length (inches)* | *Signal at Receiver (dBm)* |
|---|---|---|
| Can | 19 | –80 |
| Can | 25 | –80 |
| Can with radials | 19 | –78 |

Photos by the author.

ARRL member and Amateur Extra class licensee John Portune, W6NBC, received a BS in physics from Oregon State University in 1960 and a BA in liberal arts and communications from Ambassador College in Pasadena, California in 1963. He has also earned an FCC Commercial General Radiotelephone license and an FCC radiotelegraph license. John retired as a broadcast television engineer and technical instructor at KNBC in Burbank and then from Sony Electronics in San Jose, California.

You can reach John at 1095 W. McCoy Lane #99, Santa Maria, CA 93455 or at **jportune@aol.com**.

# A Ham Gentleman's VHF/ UHF Walking Stick Antenna

*Walk about freely with better coverage at a parade, on a stroll, or hang the antenna in a tree. Its utility will impress you.*

John Portune, W6NBC and Ernie Sloan, W6ND

**Figure 1 — John, W6NBC, holding the gentleman ham's walking stick in an operating position and as a walking stick.**

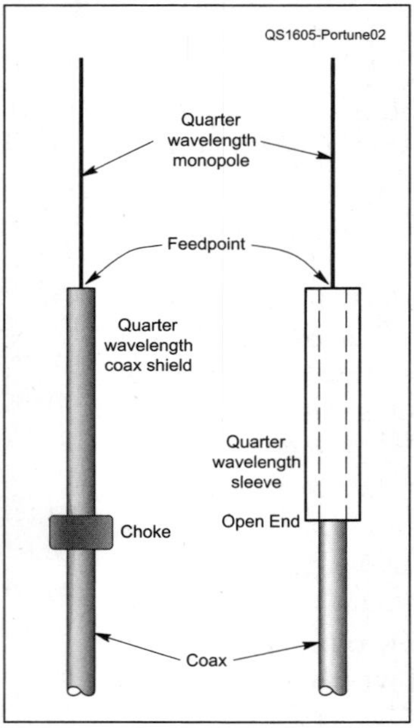


**Figure 2 — A conventional coaxial dipole (right), and a choke dipole (left).**

Over the years, walking sticks have held stilettos, whiskey flasks, fencing foils, and even guns. This walking stick conceals an efficient dual-band VHF/UHF antenna that will greatly extend the range of your handheld transceiver.

VHF/UHF portable antennas are often seen clamped to a picnic table, mounted on a tripod, or hung from a tree. But at a recent parade, it struck us that a more versatile design might be a classic gentleman's walking stick antenna (see **Figure 1**). Hold it up when you want good range, or use it as a walking stick. Our design evolved from the coaxial dipole.

## The Choke Coaxial Dipole

The coaxial dipole design consists of a $\lambda/4$ monopole mounted over a $\lambda/4$ cylindrical metal sleeve (see **Figure 2**, right). The feed coax enters through the bottom end of the sleeve and runs up to the center of the

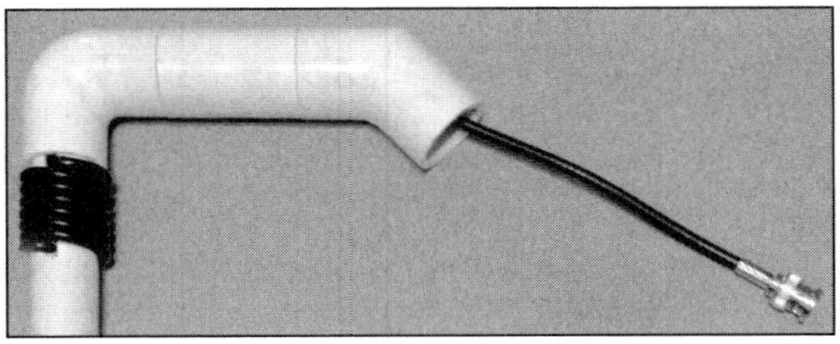

Figure 3 — The handle of the walking stick shows the seven-turn choke and the feed line pigtail. You can mount a connector instead, if you prefer.

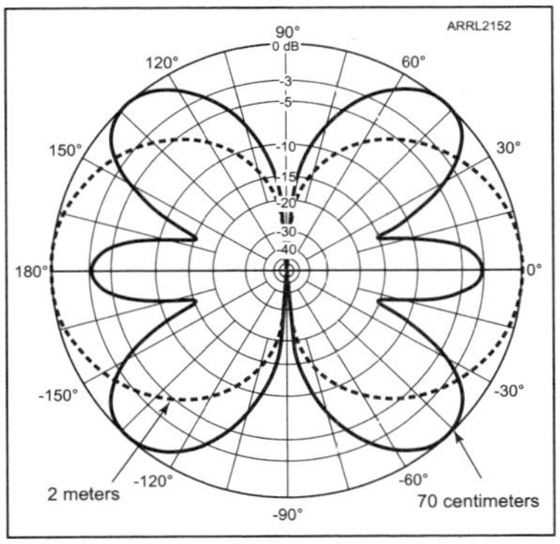


Figure 4 — The 2 meter dipole vertical free space radiation pattern (dotted) and the same antenna at 70 centimeters (solid) without stubs. The two patterns are nearly identical after stubs are installed.

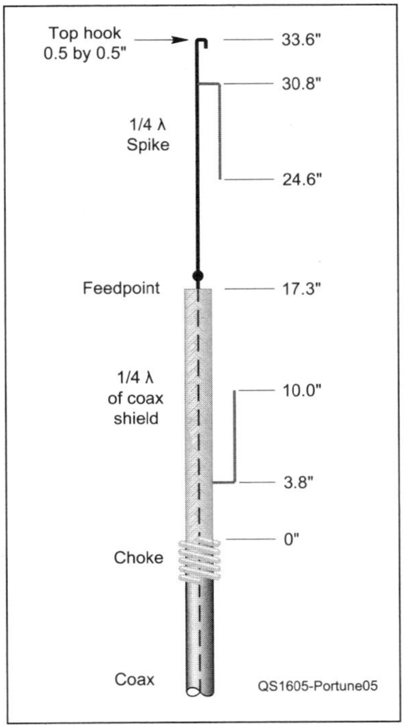


Figure 5 — Dimensions of the antenna (not to scale). Stub center-to-center width is 0.5 inches. The top hook is 0.5 inches on a side.

antenna. The inner conductor connects to the monopole and the shield to the sleeve. Our coaxial dipole design (see Figure 2, left) avoids the coaxial sleeve.

When RF arrives at the center feed point, it runs upward on the monopole, and also downward on the outside of the shield. However, we placed a choke a $\lambda/4$ down the coax where the open end of the sleeve of a conventional coaxial dipole would be. That blocks the currents going any farther down the shield. The choke is seven turns (roughly 1.5 $\mu$H) of the feed coax wrapped around the outside of the walking stick, just below the handle (see **Figure 3**).

## Dual-Banding the Antenna

Although a 2 meter dipole antenna can exhibit a low SWR on 70 centimeters (a near third harmonic of 2 meters) the radiation pattern in the 70 centimeter band is poor (see **Figure 4**). We installed 70 centimeter stubs on the 2 meter antenna to create a UHF radiation pattern that closely resembles the VHF pattern.

We placed two 70 centimeter stubs, one a $\lambda/4$ above and the other a $\lambda/4$ below the feed point, so that the outer ends of the antenna "disappear" on 70 centimeters and the radiation pattern becomes ideal. With the stubs installed, the antenna is a $\lambda/2$ dipole on 70 centimeters. On 2 meters, the stubs merely add loading, shortening the overall antenna by roughly 3 to 5 inches, but have no effect on the 2 meter pattern.

Our design dimensions (see **Figure 5**) are a compromise because the 2 meter and 70 cen-

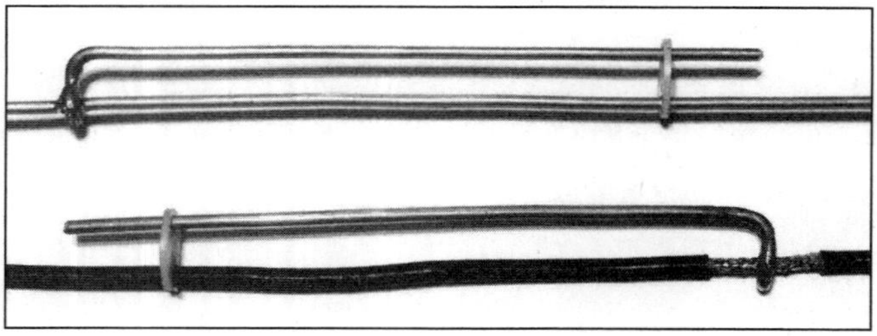

**Figure 6 —** 70 centimeter stubs with drilled nylon support washers for spacing and stability.

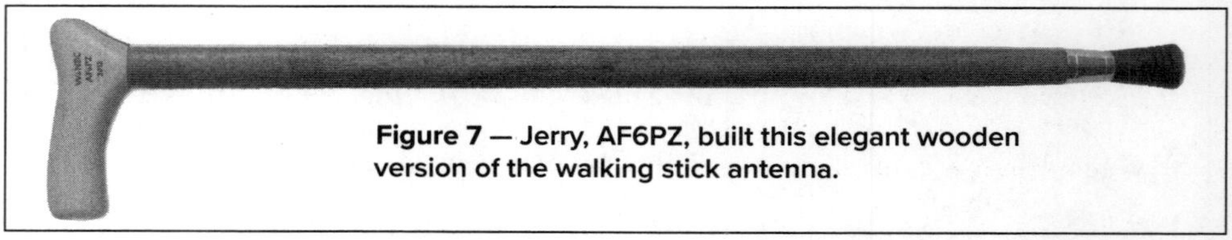

**Figure 7 —** Jerry, AF6PZ, built this elegant wooden version of the walking stick antenna.

timeter bands are not quite in an exact 3 to 1 frequency ratio. We achieved ~1.6:1 SWR and 1.2:1 at the 2 meter band edges, favoring the upper part of the band. The SWR was 1.4:1 at 440 MHz and about 2.2:1 at 450 MHz. The resulting minimum SWR is just a little high in frequency on 2 meters and low in frequency on 70 centimeters.

## Mechanical Considerations

Our walking stick is made from every ham's favorite material — common ¾-inch PVC pipe. Cut a ½-inch by 2½-inch slot lengthwise in the top end of the shaft. A Dremel® tool with a cutoff disk works well here. The slot facilitates bringing the coax of the choke to the outside of the PVC (see Figure 3). The top turn of the choke is secured by the elbow of the handle. Don't cement this joint until after final tuning.

For a permanent installation, you can use a straight PVC coupling and add length to the pipe instead of forming the handle. Now the antenna can mount quickly on a PVC mast for fixed use.

## Building the Antenna

Prepare the two UHF stubs from #8 AWG solid copper wire. Do not substitute a different wire thickness or type. The stubs, the spike, and the shield of the coax should be nearly the same diameter for a low SWR on both bands. #8 AWG wire is roughly the same diameter as the shield of RG-58 (not including the jacket).

With a pair of pointed-nose pliers, prepare the stubs as shown in **Figure 6.** Begin with wires longer than needed. The small loops at the stub ends should fit snugly on the spike (top) as well as around the coax shield (bottom) with the outer jacket removed. Ensure that both stubs have a center to center spacing of ½ inch.

Next, prepare the RG-58 coax for the lower half of the antenna and the coax pigtail. Carefully remove the outer jacket for about 1 inch where the lower stub solders to the shield. Also, leave roughly 1 inch of the center conductor sticking out of the top of the coax above the feed point. You will later solder this to the bottom of the spike, 0.3 inches above the feed point.

Next, cut the spike to 17 inches in length, and bend a 0.5 by 0.5 inch hook at the top end. The spike will attach at the dot 0.3 inches above the feed point in Figure 5. Now solder the stubs to the antenna halves at the positions indicated. Tin all connection points before soldering. Apply your iron only to the small loop at the end of the stub to avoid overheating the coax, not directly to the coax shield. When the solder flows, it will bond to the shield. Then cut the stubs to length.

Nylon washers with a ¾-inch OD, and ¼-inch ID provide spacing and stability for the stubs. Drill opposing holes ½ inch apart in the washers for the wires and the coax (see Figure 6). Finally, solder the two halves of the antenna together at the dot in Figure 2. Form a small loop in the center conductor 0.3 inches above the feed point. Stabilize this connection with heat-shrink tubing or electrical tape. The antenna may otherwise break at this point during assembly or later during use.

To assist during tuning, measure 3.8 inches down the coax from the bottom of the lower stub, and put a small tie-wrap around the coax just above it. Now push the coax through the handle and insert the completed antenna assembly into the walking stick shaft from the slotted end. Next, pull the coax to the outside through the top of the slot, and snug it up against the tie-wrap. Wind seven closely-spaced turns onto the shaft, returning the coax to the inside of the shaft at the top as seen in Figure 3.

While helping the coax through the handle, push the handle onto the shaft. Do not yet cement this joint. For long-term stability in a fixed installation, you may also wish to install a nylon screw through the shaft to support the top hook of the antenna assembly inside the PVC pipe.

## Tuning

The only tuning required is to set the distance between the lower stub and the choke for a 2 meter SWR dip at 147 MHz. While tuning, it is important to suspend the antenna clear of nearby objects with a length of non-metallic cord.

If the resonant frequency on 2 meters is low, the choke is too far from the lower stub. Pull off the handle and slide the tie-wrap and rewind the choke windings closer to the stub. If it's too high, reverse the process. The resonance on UHF is determined mostly by stub length, and is little affected by the location of the choke. After final tuning, cement the handle onto the shaft. You may choose to weatherproof the choke with large diameter heat-shrink tubing or electrical tape, and to add a PVC pipe cap to the cane end opposite the choke.

Many thanks to my co-author Ernie, W6ND, for inspiring this project with his own version of the walking stick.

## Adding a Touch of Elegance

After seeing our original PVC canes, Jerry, AF6PZ, a woodworker, crafted a magnificent version of it (see **Figure 7**) with an SMA connector neatly fitted into the handle. The details of its construction are beyond the scope of this article, but we'll be happy to e-mail readers a full description by AF6PZ.

---

Photos courtesy of the authors.

John Portune, W6NBC, is a native of Los Angeles, California. He has been licensed for 51 years, and as Amateur Extra class since 1972. He holds a British call sign, MØGCK, and commercial General Radio Operator and Second Class Radio Telegraph Licenses. He has a Bachelor's degree in Physics from Oregon State University. John retired from a career as a TV broadcast television engineer/instructor, for KNBC in Burbank, California, and Sony Broadcast, of San Francisco. He has published in *WorldRadio*, *73*, and *QST*, winning the Cover Plaque Award in October 2009. John is active in many modes on HF through UHF including satellites, and low and medium frequency experiments, and he is a Volunteer Examiner team leader. John speaks frequently at local Amateur Radio clubs, and is past president, Satellite ARC, Vandenberg Air Force Base. He is also a steam railroading and pipe organ enthusiast. You can reach him at **jportune@aol.com**, or through **w6nbc.com**.

Ernie Sloan, W6ND, is a native of Oklahoma, and moved to California in 1962. He has been licensed for 60 years. Ernie retired from management in aerospace at Vandenberg Air Force Base, and from field engineer duties in Europe. He is a member of the Satellite Amateur Radio Club, Vandenberg AFB. He is active on phone, CW, and digital modes on HF, VHF, and UHF, and he has worked the International Space Station and satellites. Ernie is an active DXer, top of the Honor Roll, and has earned WAS. He is a Volunteer Examiner. Ernie is net control for one of the largest 2 meter nets on the west coast. His antenna systems are all homebrew or modified. You can reach Ernie at **w6nd@arrl.net**.

# An Efficient 2 Meter Antenna Disguised as a TV Satellite Dish

*This vertically polarized horizontal slot antenna, cut into the reflector of a TV dish, might be the ultimate stealth antenna.*

John Portune, W6NBC

I've long wondered if it would be practical to hide an efficient 2 meter base-station antenna in a TV satellite dish. My homeowners covenants, conditions, and restrictions (CC&R) committee couldn't force me take it down.[1] But would all the metal in the TV dish compromise a 2 meter antenna? Finally, it struck me: don't fight the metal, take advantage of it, by cutting a half-wavelength-long slot antenna into the TV dish reflector.

A slot antenna is a narrow rectangular opening in a large conductive surface, such as a TV satellite dish. Slot antennas are familiar in the commercial radio world. They're common in TV broadcasting, the skin of aircraft, and in radar, microwave, and cell phone applications. This TV dish (see **Figure 1**) slot is the complement to a dipole. It is also a great way to learn about slot antennas.

**Figure 1 — Note the low visibility of the slot in my modified satellite TV dish.**

## The Slot Antenna

A slot behaves like a dipole. Both have roughly 2 dBi gain perpendicular to the antenna and they are omni-directional on axis. I found that a slot is easy to work with. I was delighted to find that this one displays good agreement with theory. The width of a slot — like the conductor diameter of a wire dipole — partially determines its bandwidth. My first slot was arbitrarily ½ inch in width, but all smaller widths I subsequently tried also worked well. The one described here is a single jigsaw blade cut. I reasoned that the thinner the slot, the less neighbors would notice it. **Figure 2** graphs its very adequate bandwidth — less than 1.5 to 1 SWR over the entire 2 meter band.

There are differences between a slot dipole and a wire dipole. The horizontal slot is vertically

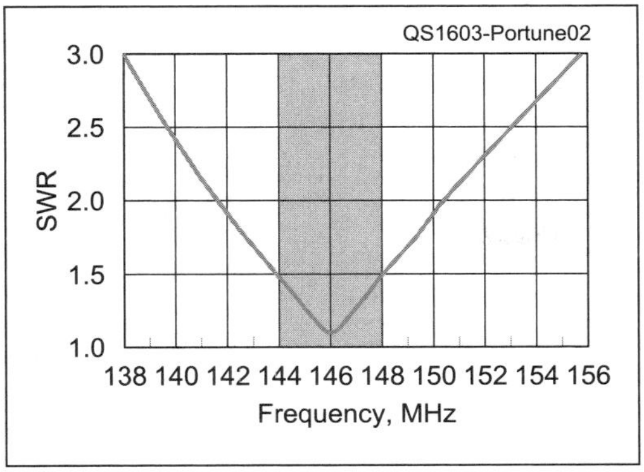


**Figure 2 — Bandwidth of a single-saw-blade-cut TV dish slot antenna.**

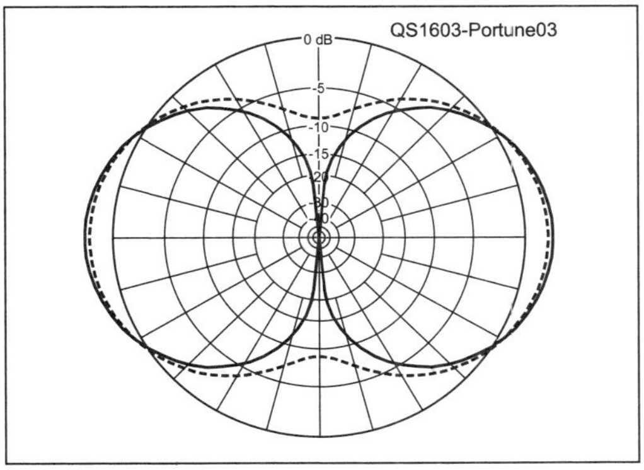


**Figure 3 — Elevation radiation patterns for a vertical wire simulation of the bent slot (dotted) and a straight slot (solid). The peak gains are 1.9 dBi and 2.1 dBi, respectively.**

polarized because of the RF currents flowing in the entire surface around the slot. The horizontal dish slot is correct for vertically polarized 2 meter repeaters.

## Dish Size and Slot Size

The common 22 × 32-inch SlimLine DIRECTV® dish, or equivalent, is a good choice. It has a large enough surface area. My *Antennas* textbook suggests that classical slot behavior is achieved for a surface larger than ½ λ by ¾ λ.[2] This made me wonder if even a SlimLine dish would be too small, but tests suggested otherwise. This slot behaves classically. I welcome reader comments on their experiences.

There isn't enough room for a straight slot in a SlimLine DIRECTV dish — the ends must droop in order for the slot to fit. This only slightly "softens" the gain and directivity. **Figure 3** shows the comparison between wire equivalents of the slot, and of a dipole. I used wire equivalents for the simulations because *EZNEC* is not ideal for slots.

## Cutting the Slot

To cut the slot, remove the reflector from the mount and LNB assembly. Mark a straight 24-inch horizontal line and two 7.5-inch drooped ends (39 inches total) on the back. Position them as shown in **Figure 4**. Exact position is not critical. Make a single jigsaw cut with a 21 tpi HSS metal-cutting blade. File the edges smooth and apply clear plastic spray paint to prevent rust. Do not cover the slot. The neighbors won't notice the slot at this width.

My 40-inch-long cut is a little too long. I did this intentionally to cause the slot to tune a little low in frequency. A too-short slot would have been difficult to lengthen with the dish mounted in its location. It's much easier, and just as effective, to "short" a slot end with heavy aluminum tape (a hardware store item). If the short is large compared to the slot (roughly 1.5 × 1.5 inches), capacitive coupling makes it unnecessary to remove the paint from the dish.

For stability, the dish needs a rigid plastic brace across the slot (back center). I used 3 inches of 0.5-inch square nylon bar stock

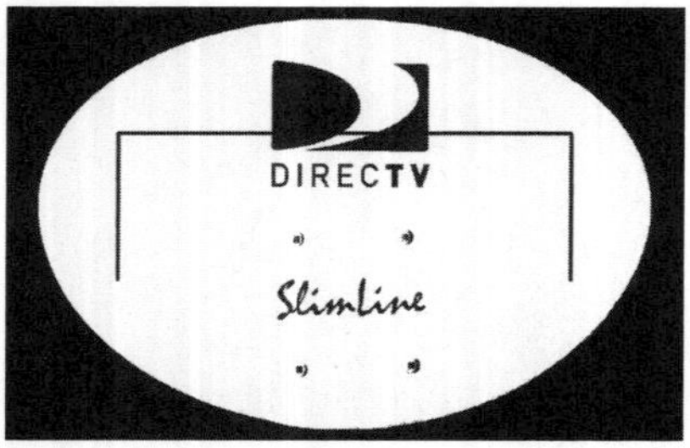


**Figure 4** — A horizontal slot antenna, with drooping ends, cut into the reflector of a SlimLine DIRECTV satellite dish.

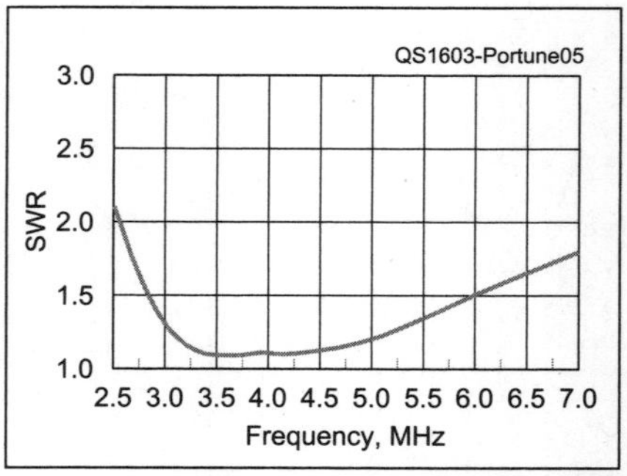


▲**Figure 5** — SWR for a feed point approximately 4.5 inches from the end of the slot.

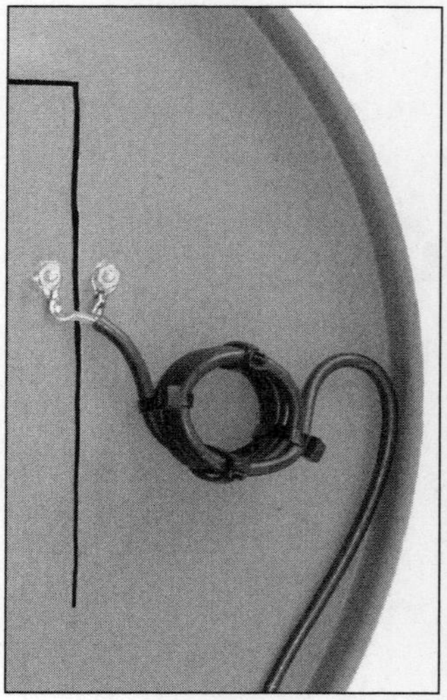

▶**Figure 6** — The coax feed line includes a six-turn coiled-coax choke balun and attaches 4.5 inches from the slot end.

attached with four 6-32 brass screws. The plastic handle of an old paint brush, for example, would also be fine.

## Feeding and Matching

Attach the feed coax braid to one side of the slot and the center conductor to the other side directly across, though not in the middle. Coax feed for a slot is done off-center. A center-fed wire dipole impedance is $Z_{dipole} = 72\ \Omega$, increasing off-center toward the ends. Conversely, the slot dipole impedance is high in the middle ($Z_{slot} = 493\ \Omega$), and decreases toward the ends. My *Antennas* textbook estimates the 50 Ω points at roughly 0.05 λ from either end. [The wire dipole and slot dipole are complementary structures. Their center-fed impedances are related by $(376.7)^2 = 4Z_{dipole}Z_{slot}$ and their electric and magnetic fields, and thus polarizations, are swapped. — *Ed.*] Using my MFJ-259B antenna analyzer, I easily found the match near the textbook estimate.

I initially used an easy-to-move feed-line attachment fixture and found that the feed point location is not touchy (see **Figure 5**). You needn't duplicate my movable fixture — just directly make a permanent attachment as in **Figure 6**.

The match is good anywhere from 3 to 6 inches from an end. Be sure to use a 1:1 current choke balun. The one shown is six turns of RG-58 coax secured with UV-stabilized Ty-Rap® cable ties. I used clear silicon adhesive to attach the choke to the back of the dish.

## Using the Dish

After seeing my dish, my ham friends asked if one dish could be used for both 2 meter activity and TV reception. Possibly, but recalling that the slot has azimuth directivity, you might want to point your slotted dish at the repeater(s). However, it's omni-directional in elevation, so you can tilt it upward as if it were receiving TV satellite signals. For a good path to 2 meter repeaters, mount the dish at least at roof height, as you would any 2 meter antenna.

Dual operation may not be practical because 2 meter transmissions might interfere with TV reception. I have not tested this, and welcome reader comments. I prefer separate dishes, on opposite sides of my house.

## Finishing the Dish

I left the TV LNB assembly as-is. The arm is perpendicular to the slot and the whole assembly is likely not resonant. With a little effort, though, a false non-metallic feed assembly could be fabricated. I did not consider the effort worthwhile.

With one of these disguised dishes, next time you're talking on your local repeater you can say, "I'm coming to you through my TV satellite dish." That should spice up the channel chatter.

**Notes**

[1]US 47 CFR Section 1.4000, October 1996, "prohibits restrictions that impair...antennas used to receive video... including...satellite dishes less than one meter (39 in.) in diameter." **fcc.gov/guides/over-air-reception-devices-rule**

[2]J. Kraus, *Antennas*, 3rd Edition, Chapter 9, p 305.

---

John Portune, W6NBC, can be reached at 519 W Taylor St, SPC 111, Santa Maria, CA 93458, or by e-mail at **jportune@aol.com**. His website is **w6nbc.com**.

# The Rooster-Tenna – A 2-Meter Weathervane Antenna

*Your neighbors will love the rooster and you'll love the performance.*

John Portune, W6NBC

Weathervanes are great for making an antenna look like it belongs on a rooftop to avoid receiving complaints. For instance, my weathervane just happens to also be a high-performance 2-meter antenna (see the lead photo). A Rooster-Tenna is a great disguise.

In the August 2005 issue of *QST*, I presented an earlier rooster-vane antenna. This new version is much improved. The antenna is now a wide-spaced, parallel-fed horizontal folded dipole bent into a flattened cube. Technically, it is a skeleton slot. It's built around two horizontal square aluminum tubing rings, 11½ inches on a side, separated by 5½ inches (center-to-center), and electrically connected by three vertical interconnect sections.

The rings are bent from rigid ½-inch OD aluminum tubing. Hardware-store soft copper or aluminum tubing is not as durable. Local metals dealers, **DXengineering.com**, **metalsdepot.com**, and **onlinemetals.com** all stock it. At the corners, the rings are held apart by four ½ × 5 inch square acrylic posts, which also mount the vane's N-S-E-W letters. The posts are tapped lengthwise for 10-24 screws through the rings. The mast is 1-inch thin-wall (.063-inch) aluminum tubing. Only the size and spacing of the rings need to be what is shown in **Figure 1**. The other dimensions are just suggestions. **Table 1** shows the list of materials.

The vertical sections that connect the rings are made from ¼-inch soft aluminum tubing. The plain ones, left and center in **Figure 2**, are the ends of the folded dipole. You adjust the antenna-matching SWR by moving the vertical section on the right, the one that contains the feed point. Adjust the operating frequency by moving the section on the left.

## Performance

**Figure 3** shows that the antenna SWR is less than 2.5:1 across the entire 2-meter band. **Figure 4** shows the free-space *EZNEC* eleva-

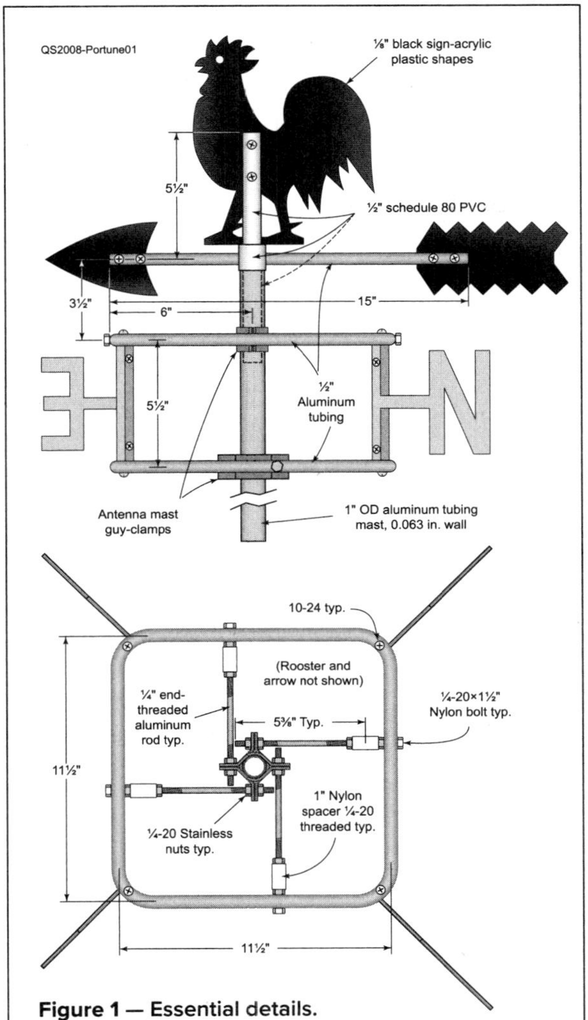

Figure 1 — Essential details.

tion, azimuth, and 3D radiation patterns with gains in dBi. For a comparison, a half-wavelength vertical dipole or J-pole has an azimuth gain of roughly 2.1 dBi. The Rooster-Tenna is more nearly isotropic, so it has only a little less horizontal gain but higher overhead gain, which can be useful for satellite work. The antenna is omnidirectional in azimuth, and as a horizontal slot, it is vertically polarized — best for accessing repeaters and mobile stations.

Figure 2 — Vertical interconnect sections, the frequency adjustment at left, and the SWR adjustment at the feed point at right.

## Mounting

Four ¼-inch partly threaded aluminum rods — two for each ring at 90 degrees — mount the antenna on the mast. The rods attach to the mast by two nuts at the ends of common TV-type antenna guy-cable clamps, available from **RadioShack.com** online, catalog number 1500335, and many local electronics parts stores. One-inch threaded nylon spacers and nylon bolts insulate the rods from the antenna rings. You will need to thread 1½-inch and 1-inch of the rod ends, respectively, with ¼-20 threads. I used an inexpensive common hardware-store threading die. Aluminum all-thread rod may also be used.

## The Rotating Rooster and Arrow

The top assembly, with the rotating directional arrow and rooster, is a separate unit that turns inside the 1-inch thin-wall (.063-inch) aluminum tubing mast. The inside diameter of the mast is a close fit for the ½-inch PVC water pipe that mounts the rooster and arrow. White Schedule 40 PVC is acceptable; grey Schedule 80 PVC is stronger and less visible.

Cement the upper and lower PVC pipe sections into a ½-inch, PVC coupling and drill a ½-inch through-hole from the side for the

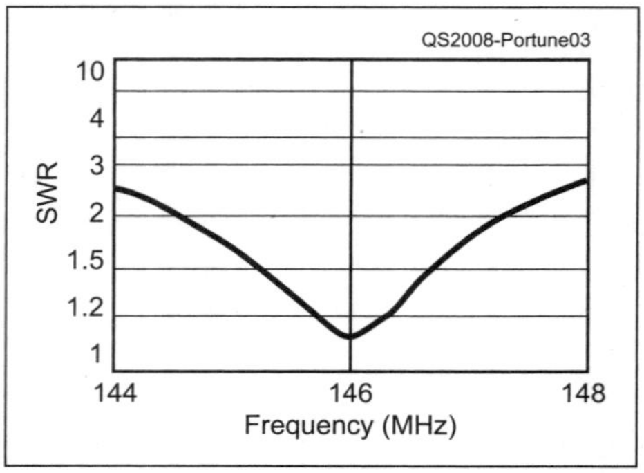


**Figure 3 — The SWR from 144 to 148 MHz, measured with a RigExpert A1400.**

**Table 1: Materials**

| Item | Source |
|---|---|
| 10 feet of rigid ½-inch OD aluminum tubing | Local or online metals dealer |
| 18 inches of ¼-inch soft aluminum tubing | Hardware source |
| Mast: 1-inch OD thin-wall (.063-inch) aluminum tubing | Online antenna tubing supplier |
| TV antenna guy-cable mast clamps (2 count) | TV antenna hardware source |
| 6-foot coax pigtail with connector | Electronics source |
| 6-32 stainless-steel screws and nuts | Hardware source |
| 10-24 stainless-steel screws | Hardware source |
| ¼-20 stainless-steel hex nuts | Hardware source |
| ¼-20 nylon bolts and nuts | Hardware source |
| 1-inch nylon spaces, ¼-20 threaded | Hardware source |
| Plastic parts | Craft store or sign-making shop, and 3D printer files at **arrl.org/qst-in-depth** |
| 20 inches of ½-inch-square acrylic or other plastic rod | Internet plastics source |
| 30 inches of ¼-inch aluminum rod | Bicycle safety flag mast |
| Tie-wraps (UV stable variety) | Hardware/electronics source |
| 1:1 choke balun, ferrite, or coax | Hardware/electronics store, and 3D printer files at **arrl.org/qst-in-depth** |
| ⅜-inch stainless-steel cable clamps (6 count) | Hardware/electronics store |

arrow. A second 6-32 screw and nut at 90 degrees secures the arrow. The bottom PVC pipe section should be roughly 12 inches long so that the top assembly will stay in the mast under high wind conditions. Cut a ⅛-inch slot in the top end of the top pipe for mounting the rooster, using 6-32 screws and nuts. Do the same in the ends of the arrow shaft for the arrowhead and feathers. Use stainless-steel or brass fasteners throughout this project.

## Making the Antenna Rings

I used a tubing bender (see **Figure 5**) available on **amazon.com** that makes precise 1½-inch-radius bends in ½-inch tubing. This tool has proven handy for several new antenna designs. Similar models are available for other tubing sizes. A less-expensive method, suitable for soft aluminum tubing, is to use bending springs. They are readily available on the internet and from local hardware stores. A short length of larger-diameter pipe, held in a vise, provides a bending mandrill for use with the springs.

Alternately, you may make the rings as four separate straight sections. Flatten the ends and drill holes for 10-24 screws, ⅜ inch from the ends. The screws that attach the acrylic posts and N-S-E-W letters will now connect the four ring sections together. The rings should be 11½ inches on a side, tubing measured center to center. The tubing ends should meet in the middle of one of the ring sides. Connect the ends with a short metal link or heavy wire, secured by screws and nuts.

## Making the Interconnect Sections

Cut three 4¾-inch lengths of ¼-inch soft aluminum tubing (see Figure 6). Flatten ¾ inch at both ends. Be careful to flatten the ends in the same plane. Then drill ⁹⁄₆₄-inch holes ⅜ inch from the ends for 6-32 screws.

Next, remove ⅜ inch from the middle of one of the sections for the feed point and flatten the ends. Fabricate a small plastic insulator for the gap and secure it with 6-32 screws and nuts. These screws become the connection terminals for the feed coax.

Rubber-insulated stainless-steel cable clamps provide an easy way to attach the vertical sections to the antenna rings. They also permit adjustment of tuning and matching. Online sources, as well as many local electronic parts stores, offer these clamps. They are, however, a little large to create a good connection to the rings. Therefore, leave roughly ⅜ inch of the rubber insulation on the clamp to ensure a tight connection.

## Plastic Antenna Hardware

If you have access to a 3D printer, a complete set of .stl files for printed plastic parts is on the *QST* in Depth web page at **arrl.org/qst-in-depth**. ASA filament is best for its UV resistance. The antenna uses conventional home workshop construction methods and materials. Templates for hand cutting the plastic parts are also available on the *QST* in Depth web page, along with *AutoCAD* .dwg files. From these files, a local sign-making shop was able to laser cut the letters, rooster, and arrow parts at a modest cost. Sign acrylic plastic is an excellent material, as it is nearly totally impervious to the UV in sunlight.

Alternately, ready-cut acrylic letters are available on the internet as well as from some craft stores. A further possibility is to cut the shapes from the flat sides of a large rectangular plastic trash can. These are generally made of heavy polyethylene plastic, which is also UV resistant and easily cut with tin snips.

## Feed Coax and Balun

For the feed point, separate roughly 2 inches of the center conductor and braid of a 6-foot-long RG-58 or Mini-8 coax pigtail. Weatherproof the ends with heat-shrink tubing. Crimp on #6 ring terminals for the feed point screws. Add a connecter at the other end. Beyond the pigtail, it is best to use larger coax to avoid the high loss of small-diameter coax at VHF.

You should provide the antenna with a 1:1 current choke balun. It may be a small stack of VHF mix ferrite beads on the coax pigtail or a small 4 – 6 turn coil wound with the coax of the

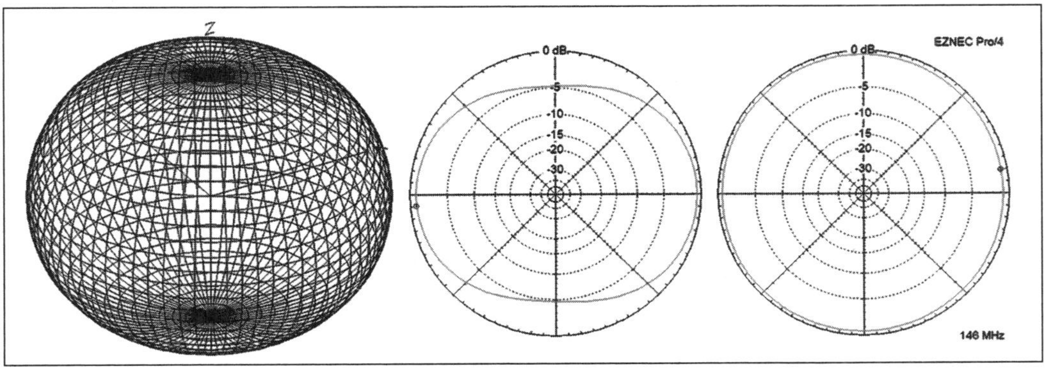


**Figure 4** — From left to right is a free-space 3D radiation pattern, vertical elevation pattern, and horizontal azimuth pattern. The gains are in dBi.

pigtail, bundled and secured with zip-ties. Attach the balun to the mast with zip-ties just below the antenna. Included also for 3D printing is a quick-form for a coax balun using small-diameter coax (see **Figure 7**).

## Tuning and Matching

Begin adjustment with the feed point section roughly 1 inch from the middle section and the other roughly 6 inches away. The three sections may also be moved as a group, if required during tuning. For better stealth, the feed point section is best placed near one of the fiberglass supporting rods on the bottom ring. Zip-tie the coax to the rod and main mast.

Adjust the SWR first using an antenna bridge or analyzer. Once the SWR is low, it will change only a little when the frequency is adjusted. Therefore, pay no attention to frequency until you obtain a low SWR. Fine tune both with the antenna mounted in its permanent operating location.

When neighbors or homeowners associations first see this attractive antenna, they are not likely to notice it or see it as unwelcome.

John Portune, W6NBC, is an ARRL member and frequent contributor to *QST*. He has been licensed for 54 years and has held an Amateur Extra-class license since 1972. John has a BS in physics and also holds an FCC Commercial General Radiotelephone Operator license and an FCC Radiotelegraph license. He retired as a broadcast television engineer and technical instructor at KNBC in Burbank and then from Sony Electronics in San Jose, California. You can reach John at **jportune@aol.com** or through his website at **w6nbc.com**.

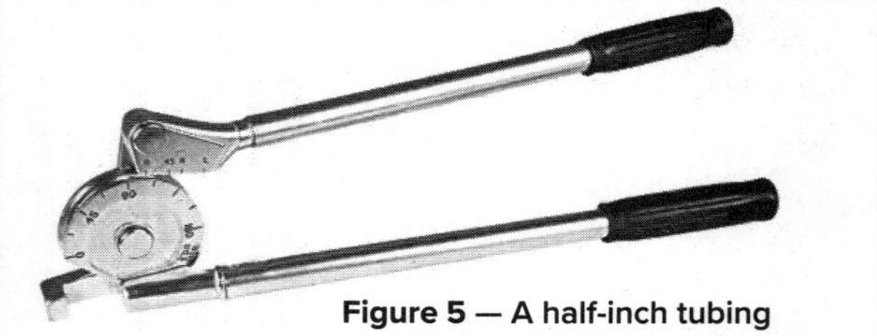

**Figure 5 — A half-inch tubing bender for 3-inch radius bends.**

**Figure 7 — 3D-printed quick-form for a VHF balun using RG-58 or Mini-8 coax.**

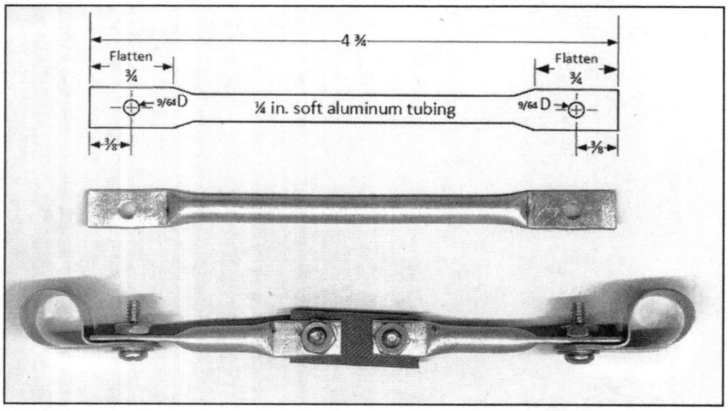


**Figure 6 — Vertical interconnect sections. The lower section is shown with clamps. Note the plastic insulator in the lower feed point section.**

# The Fountain Antenna

***This multiband antenna for portable or home use is easy to build and operate.***

Ted Coyle, AA4AZ

**The antenna elements and locations where they're secured. All elements are quarter wavelength. (Elements from left to right are for 60, 40, 20, and 30 meters.)**

I recently decided to build an HF antenna that would let me simulate emergency conditions during both the build process and operation of the antenna. As I created the design, I assumed that the operator had no access to an existing antenna and needed to create one. My design goals included:

- Simple construction; relatively quick to build and deploy
- Use of materials that could be easily found
- Capability of portable or semi-permanent use as needed with minimal effort
- Reliable operation with little or no need to make physical adjustments for tuning or band changes
- Multiband capability without the use of tuners, traps, switches, etc.

I decided to build an HF multiband vertical with resonant elements for each band and a radial field that was relatively small for portable or semi-permanent use as needed.

## The Fountain Antenna

The final design uses a common feed point on the ground with vertical and sloping elements going in different directions and secured to different points (see the lead photo). Viewed from the side, it suggests the appearance of individual streams of water coming out of a fountain. Each element is supported by a tree branch from a different nearby tree. The design can easily be duplicated as long as there are trees nearby.

I originally planned for 40 meters as the longest element, but I found I could also add an element for 60 meters. The performance was acceptable even with a slope of nearly 45 degrees and no changes to the radial field.

The final design has separate quarter-wave elements for each of the 60-, 40-, 30-, and 20-meter bands. As a bonus, I also get 15 and 6 meters, making this antenna operational on six bands in total. Changing bands requires no effort, no tuning, and no switches. As with a fan dipole, the RF picks the path to the resonant element.

For the length of each quarter-wave element, I used the formula "length in feet = 234/frequency in MHz," cutting a little long to allow some trimming with my antenna analyzer. The formula numbers were close to the final lengths, meaning it should give good results even if an analyzer is not available. Interaction between up to four elements is minimal due to the separation of the elements. Tests with more than four elements led to difficult tuning due to interactions.

The feed point is secured to the ground by small bungee cords attached to two stakes (see **Figure 1**). The 40- and 15-meter element is vertical. Each of the other elements slope at various angles relative to the ground, and in different directions. The 60-meter element is the longest and is at about 50 degrees, due to the difficulty in securing it to a higher branch. The 30-meter element is at about 70 degrees, and the 20-meter element is at around 80 degrees. Despite the sloping elements, hundreds of contacts have shown that the patterns are nearly omnidirectional.

There is no magic in any of these angles or directions. I decided which bands I wanted, put the feed point where I had a clear spot, and located each element where I could find the best fit. The configuration will be different for each installation due to the unique conditions at each location. Again, this build is meant to simulate an emergency situation where the objective is to get on the air relatively quickly.

## Cost

Total cost for the final multiband antenna using elements, radials, feed point, support lines, and connectors was about $25, not including the coax. The elements and radials are made of #16 AWG stranded, insulated wire from a surplus store. Other materials include a PVC box, stainless-steel hardware, support line to suspend the elements, and an SO-239 connector. Even with all new materials, this antenna could be built relatively inexpensively.

**Figure 1** — Antenna elements departing the feed point at various angles. (Elements from left to right are for 20, 40, 30, and 60 meters.)

**Figure 2** — Details of connections to the feed point. Note the optional use of a pigtail linking the radials to the feed point, simplifying the disconnection and reconnection of radials to the system.

A detailed parts list is available on the *QST* in Depth web page (**arrl.org/qst-in-depth**) in the **2020** dropdown.

## Feed Point

The feed point is built around a ¾-inch PVC type LR electrical conduit box, from the local home improvement store. This has a weatherproof access lid that's secured with two screws and has two openings that must be plugged, using ¾-inch PVC cap and a 2-inch length of ¾-inch PVC pipe.

On the conduit box, I mounted the SO-239 for the coax feed and two ¼-inch stainless steel bolts, with nuts, washers, and lock washers — one for the ground radials and one for the antenna elements (see **Figure 2**). Red and black PVC tape was used to indicate antenna and ground sides. The SO-239 and stainless-steel bolts are connected inside the box with two short wires (see **Figure 3**). Lugs were soldered to the feed end of each element and attached to the feed point bolt.

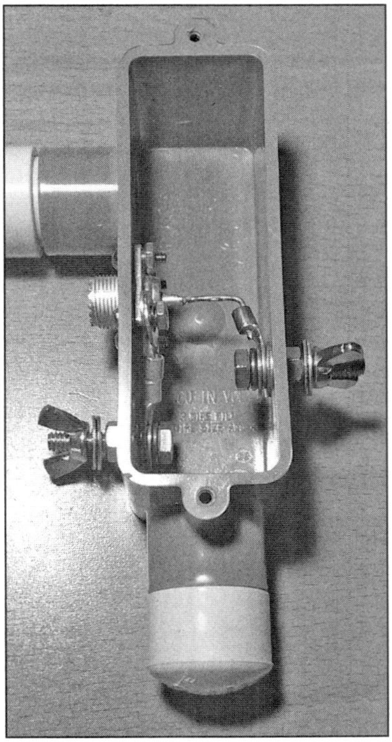

**Figure 3** — Wiring and hardware detail of conduit box feed point.

## Radials

I wanted to use the minimum number of short radials that would get me on the air quickly and reliably support communications, without needing to get every decibel of gain out of the system.

I experimented with various lengths and numbers of radials, but settled on using 16 radials, each 16 feet long. The wires are soldered into bundles of four onto each of four terminal lugs, which are attached to the ground bolt using wing nuts.

## Securing the Wires

Each radiating element and radial is secured at the end opposite of the feed point with a short length of plastic pipe hanger strap, which has holes every ¾ inch. I cut a 3- or 4-inch length for each radial and driven element, then threaded the wire through the strap material a few times.

At the other end of the plastic hanger strap, I tied the line used to suspend the element to the tree. For the radials, I inserted a large nail into the opposite end of the strap material after it was secured to the radial. I pulled the radial taut and pushed the nail into the ground by hand. This keeps the radials flat against the ground.

## SWR

The elements are cut to favor the CW portions of the bands. SWR for those portions of the bands is 1.1:1 on 60, 40, and 30 meters, and 1.3:1 on 20 meters. SWR is also respectable on the "bonus" bands of 15 and 6 meters. SWR on 15 meters is 1.2:1 or better, and on 6 meters it's 1.4:1 on the CW portion and gradually goes to 2.5:1 at 54 MHz.

## Fighting Wind

To deal with branches moving during windy days, I placed bungee cords at each end of the line that suspends any element where branch movement is a problem. Some experimentation was required in order to determine how much I needed to dampen the impact of movement while ensuring the element was sufficiently taut when the winds were calm.

## Safety

Safety must be the first consideration of this or any other antenna. This being a ground-mounted antenna, I did no climbing and used no ladders during the project. All lines were thrown over branches by hand, which limited height and resulted in a greater slope in the case of the 60-meter element, but was consistent with the simulated emergency deployment where mechanical means to launch a line higher might not be readily available.

The antenna is fairly stealthy, and can be difficult to see depending on the location where it is used. However, safety is more important than stealth, so remember that within a certain distance of the feed point, it is possible to walk into the "hot" wires that make it look like a fountain. The radials are also a potential tripping hazard. Choose a location that has little or no foot traffic, and use some sort of marking to alert people as they approach it. Limit access of people and animals to the antenna while it is in operation.

## Conclusion

I have made hundreds of two-way contacts using the fountain antenna. It is fun and inexpensive and has become my go-to antenna on the bands for which it was designed. Using 100 W, I'm pleased with the performance and the signal reports I have received.

I liked this well enough to build a second antenna, intended for portable operation. However, as band conditions improve, I plan to use it as a second semi-permanent antenna with elements for other bands and then build a third for portable operations.

---

All photos by the author.

Ted Coyle, AA4AZ, has been a licensed amateur radio operator for 52 years, and has had his Amateur Extra-class license for 47 years. He earned a BS degree from Stockton University, and a Master's of Engineering Management from Widener University. He is a retired Environmental, Health, and Safety professional, with related experience in ISO management systems and Six Sigma. He is active in the Straight Key Century Club, the Salem County NJ ARES/RACES group, and is a member of the Gloucester County, New Jersey, Amateur Radio Club. He can be reached at **aa4az@comcast.net**.

# The K4HRK ComBrella

*Have it made in the shade at your next outdoor operation.*

Bob Herklotz, K4HRK

My ComBrella project began as a novelty — a com radio and antenna integrated with an umbrella. I opened an umbrella one day and noticed that the ribs of the umbrella were similar to the drooping radials of a ¼-wave ground-plane antenna. I wondered if an umbrella could provide the ground plane for a 2-meter/70-centimeter antenna. I was excited to make it for additional utility to a traditional umbrella.

## Preparing the ComBrella

I built my ComBrella with a metal-ribbed and end-tipped umbrella, which might be difficult to find today. For this reason, I'll offer two build procedures — one for older umbrellas with metal ribs and end tips, and the other for any umbrella with non-conductive fiberglass ribs. The materials (listed in **Table 1**) are needed for either version of the build.

If you're using a metal-ribbed and tipped umbrella, before you begin, you'll need to confirm direct (not just capacitive) continuity between the metal end tip and all the metal ribs. You may need to file down the chrome or painted finish on the end tip to expose the bare metal. If you don't have continuity, you'll need to connect the coax shield to the metal ribs.

## Metal-Ribbed/Tipped ComBrella Procedure

To begin, drill a ¼-inch hole through the end tip of the umbrella using a step drill rotated by hand for better control. The hole will need to accept 6 – 8 feet of RG174 coax connected to an SMA male connector (see **Figure 1**).

Beneath the umbrella canopy and above where the rib support struts lock into place, drill a 3⁄16-inch hole at an angle into the shaft to

**Table 1 — Parts List**

| *Quantity* | *Equipment* | *Comments* |
|---|---|---|
| 1 | Umbrella | Can be either fiberglass- or metal-ribbed/tipped |
| 1 | Nagoya NA-771 | A 15.6-inch VHF/UHF antenna |
| 6 – 8 feet | Yotenko RG174 coax with SMA male/female connectors | Cut off the female connector to desired length |
| 1 | SMA female crimp con-nector | Depending on your radio, you may need a different style connector |
| As needed | RG58 coax extensions | Used to increase your antenna's height to your preference |
| 1 | PVC tubing | Will need to be tailored to your umbrella handle for mounting to a folding chair |
| 1 | De-solder wick | |
| 1 | Folding chair | Optional |
| 4 | Folding chair stakes | Optional |

accommodate pushing the coax through. Start the hole perpendicular to the shaft, then angle the drill down toward the handle of the umbrella when the hole punches through. Deburr the hole and push the coax into it at the tip, and use a scriber with a pointed hook to capture the coax and pull it through.

When the SMA male connector is almost seated at the tip, apply epoxy to the shrink wrap directly beneath the connector nut (not to the nut itself) and seat the coupling into the hole. Confirm that you have good continuity between the connector nut and the top of the end tip. For aesthetics, you can complete the connection with a piece of shrink-wrap tubing.

To ensure good continuity between the ribs — which act as radials — interweave de-soldering wick among the ribs and secure a pigtail to the umbrella shaft with a spring (see **Figure 2**). This brings the radials to a common ground. The de-soldering wick is very flexible and folds nicely when collapsing the umbrella if you leave a little slack between the ribs/radials. The braid can be taped or wire-tied to the metal shaft.

If you can't get continuity from the coax connector to the metal ribs, just remove a ½-inch section of the coax covering, and secure it to the metal shaft with a spring or wire tie. This will ensure that the braid is making good contact with the shaft.

A heat gun was used to flatten a piece of thin-wall PVC tubing to make a socket for the transceiver's belt clip. Cut this to the length to secure the clip. I used duct tape to secure the PVC to the shaft at a convenient height.

The excess coax was coiled near the support struts to form a common-mode choke, and then routed to the handheld transceiver attached to the shaft assembly near the umbrella handle (see **Figure 3**). I attached an

**◀Figure 1** — The SMA connector in the tip of the umbrella.

**▼Figure 2** — The ribs were joined using solder wick.

**Figure 3** — The handheld transceiver mounts on the shaft of the umbrella.

SMA female coupling to the end of the coax to mate with the radio. Depending on your radio, you may need an alternate connector. Finally, screw in a 2-meter/70-centimeter extended antenna into the tip of the umbrella. Take care not to over-tighten the antenna and dislodge the connector at the tip.

## Fiberglass-Ribbed ComBrella Procedure

This procedure is for an umbrella with fiberglass ribs. These types of umbrellas are widely available, making them a great option for the ComBrella. The procedure is similar to the metal-ribbed version of the project, except for loosely wrapping wires around the fiberglass ribs, soldering the wires to the de-solder braid ring, and securing the wires at the end of the ribs.

The fiberglass-tipped umbrella has a plastic end tip, so we won't make the connection to the radials there. Use the same drilling procedure above for drilling the holes in the end tip and shaft, using a small step drill rotated by hand. The hole should be beneath the canopy about a centimeter from where the shaft emerges.

Push the RG174 coax from the tip end through to the hole in the shaft beneath the canopy. You will encounter a few obstructions in the shaft, but if you wiggle and rotate the coax, you'll eventually see it at the hole below the canopy. Grab it with a hooked scriber or similar tool and pull it through. Apply epoxy to the bottom of the connector nut and seat the connector to the tip.

Next, loosely wrap any small-gauge stranded insulated wire from the de-solder braid location around each fiberglass rib to the end of the ribs, with an inch ex-tending past the ribs. One or two turns around each rib is all you need.

With the plastic umbrella shaft, you will need to connect the radials to the coax braid. Remove the covering from the coax, as detailed above, and then extend a pigtail from the de-solder braid to the exposed coax braid. Wrap it around the coax a few times and use a wire tie to secure it. The pigtail can be soldered to the coax braid; just be careful not to burn through to the dielectric. Each wire should be pulled under then over the de-solder braid closest to each rib. Then solder the wire to the braid on the side facing down toward the handle. Use heat-insulating material, such as kitchen-variety sheet silicone, to prevent burning the fabric. With the wire loosely wrapped around each rib, push the wire ends into the

**Figure 4** — The ComBrella can be folded easily for storage or transport.

holes at the end of the ribs where the fabric is threaded, leaving a little slack. When the umbrella is folded, the ribs straighten and lengthen slightly.

Snip off the ends of the radial wires, leaving about a centimeter of overhang. The holes at the end of the ribs should hold the wire with a friction fit, or you can secure the wire at the rib end with small wire ties or a dab of adhesive to each hole. To complete the build, gently screw in the antenna.

## SWR and Performance

I used a NanoVNA to measure the standing wave ratio (SWR). Both antennas gave a relatively flat reading on the VNA of 1.5:1 from 144 – 148 MHz. The SWR for 420 – 450 MHz was 1.2 – 1.8:1. When connected to my 5 W 2-meter/70-centimeter handheld radio, both bands easily hit our local Maury County Amateur Radio Club repeater 15 miles away without any elevated altitude. I received several full-quieting reports.

## Practical Applications

Our Maury County Amateur Radio Club in Columbia, Tennessee, has monitored the Chickasaw Trace bicycle races for many years at the local park, with two major races each year. I set up at our operating location with the ComBrella inserted into a PVC mast. I mounted it on a folding chair with a screw through the PVC to prevent the ComBrella handle from sliding down the mast.

We use a simplex frequency, and my signal with the net control was full quieting. From a low spot, I could hit our local repeater about 20 miles away and talk to other hams effortlessly. Gaining extra antenna height is easy by simply adding another PVC coupling, a PVC section, and a length of extension coax. My folding chair has provisions for staking it to the ground to prevent the setup from crashing down. The ComBrella's fold-up capability facilitates carrying it for field work (see **Figure 4**). Relief from the sun or rain is also a real advantage for both the radio and the operator.

## Mounting the PVC Mast

I duct-taped a short, smaller-diameter PVC section to one of the folding chair's rear legs to act as a receiver and then just slid the larger-diameter section over it. The PVC sections were chosen to provide a good fit for the umbrella handle. Dense foam spacers were used at the attachment points to allow the chair to fold normally. You may need to use a different approach, depending on your chair and the diameter of your umbrella handle.

## A Shady Design

While this may be a novel antenna design, I'm pleased with its performance and utility. If you build this folding 2-meter/70-centimeter ground-plane antenna, you'll have contacts made in the shade.

---

All photos by the author.

Amateur Extra-class operator Bob Herklotz, K4HRK, was licensed in 2014. He is a member of the Maury County Amateur Radio Club in Columbia, Tennessee. A retired auto industry maintenance planner and instrument-rated pilot, Bob is winding down his "retirement business," Fresh Aero Aviation, a manufacturer of aircraft LED lighting products and tools. He looks forward to spending more time on other pursuits, like ham radio. He can be reached at **robherk@gmail.com**.

# A 2-Meter Glove Antenna for Portable Operation

## *A five-element Yagi that folds to backpack size.*

Jim Grebey, N4KCT

I needed a 2-meter antenna that provided reliable portable operation when I was away from my shack. I wanted a solution that would be easy to travel with and offer a good level of gain. The glove antenna described here is suitable for backpacking, camping, Parks on the Air® (POTA), Summits on the Air (SOTA), and emergency operations. It makes a great addition to any go-kit and features a stealthy design that most HOAs can't object to. It is also helpful when operating with a handheld in fixed locations, such as campsites, shelters, or in my case, the back porch.

### Design Conception

The idea for this project started with a handheld radio that my wife gave me for Christmas. I have a Yaesu FT-991A with a Diamond vertical antenna for operating in my shack. I'm about 12 miles from my local repeaters, which are all south of my location. With the Diamond vertical, hitting these repeaters from my shack is not a problem. However, it's a different story when using the handheld and its stock antenna. They can reach the repeaters, but only on the fringe of their range. I also wanted to use the handheld from my back porch, but my shack is on the other side of the house, and changing over to the vertical didn't make sense. I envisioned an affordable antenna with enough gain to put me solidly into the repeaters, and one that I could use while camping. A fixed beam made sense because the repeaters I use are all in the same general direction. There are other aftermarket antennas for handhelds, but they fall short when used from more remote locations.

My solution is a five-element, 2-meter Yagi built into an antenna "glove." It provides about 10 dBi of gain, which is great for remote locations. The glove antenna uses fabric as its structural component. This eliminates the need for a rigid aluminum, wood, or PVC boom, making it lightweight and flexible enough to be rolled into a compact package (see **Figure 1**). When you are ready to deploy it

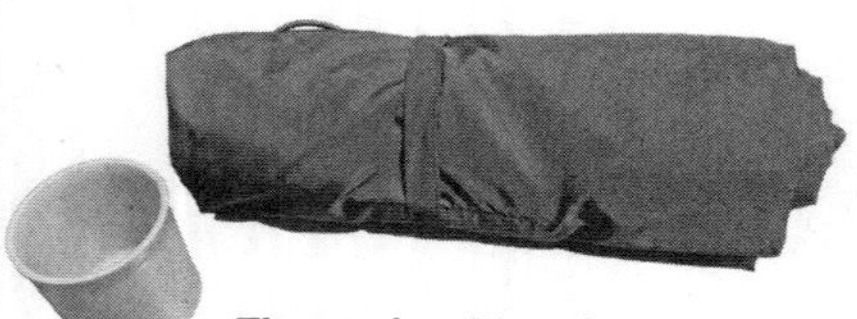

**Figure 1** — The glove antenna can be rolled into a portable package.

for portable use, it springs open and takes seconds to set up. With bungee cords, it can be hung from a tree, a porch overhang, or even an emergency shelter ceiling.

## Construction

Constructing the electrical portions of the antenna is straightforward when following the dimensions (see **Figure 2**). All of the elements can be made from a single 25-feet × 1-inch-wide steel tape measure. Apply electrical tape over the ends of each element to avoid cutting yourself or the glove. Be sure to wear safety glasses when cutting the elements from the steel tape.

When connecting the feed point, the convex side of the tape should touch the insulator, and the concave side is where you will sand off the paint to make the electrical connections. I used a high-density polyethylene (HDPE) cutting board to make the center insulator (see **Figure 3**). These cutting boards can be purchased at any kitchen supply store, and they are easy to cut and hold a self-tapping screw well. I drilled pilot holes for the screws in the insulator and in the driven elements. It is difficult to solder to the steel tape, so I used ring connectors for the 4.5-inch #14 AWG wire and for the center and shield of the coax. Use a stainless steel screw with hex lock washers to make a good mechanical bond to the driven elements. Place a second screw in each element to secure it in position.

If you have sewing skills, constructing the glove is straightforward as well. I do not know how to use a sewing machine, but my wife is an excellent seamstress (see **Figure 4**). It is important to use a lightweight yet strong fabric that will not shrink or lose its shape. I am arranging to have gloves made to order — contact me at **jim@grebey.com** if you are interested in purchasing one.

The glove is made by sewing a "finger" for

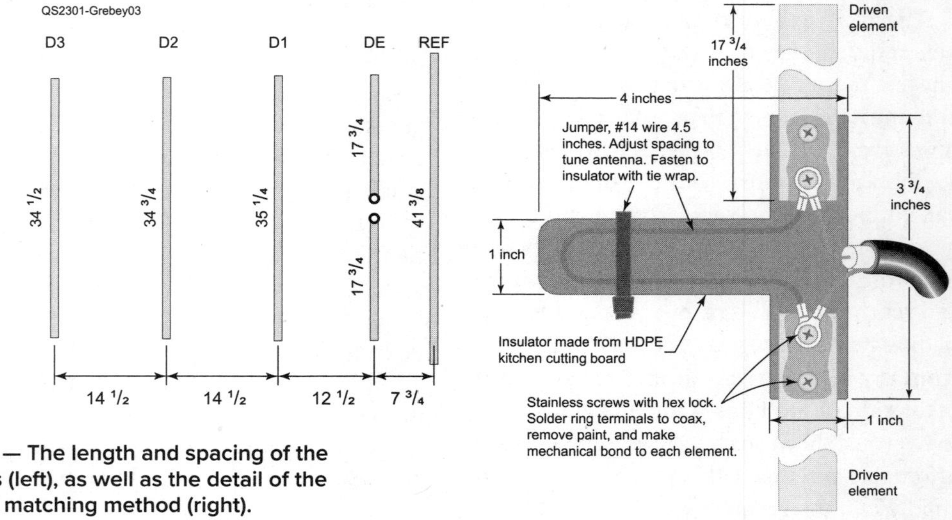


**Figure 2** — The length and spacing of the elements (left), as well as the detail of the feed and matching method (right).

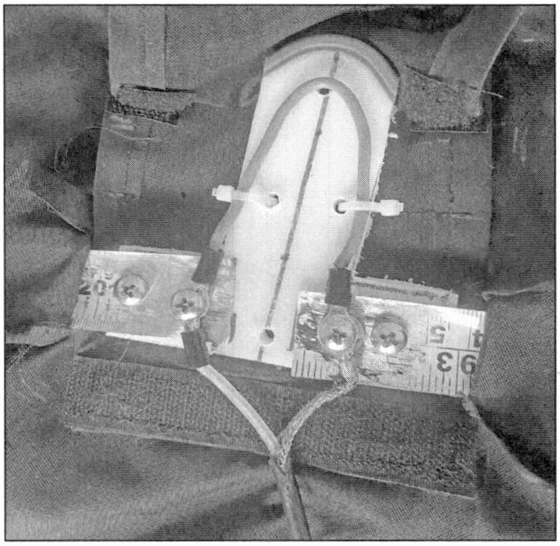


**Figure 3** — Details of the feed and match connections.

**Figure 4** — Constructing the glove portion of the antenna requires sewing skills.

each of the five elements. Each finger should be the length of the element it is holding, and just wide enough to keep the element in place. The spacing between each finger is based on the element separations shown in Figure 2. Measure from the center of the element to the center of the next element. When putting the elements in the glove, make sure the concave sides of the arches are all facing up. This will help align them when you roll up the antenna. The glove should have a grommet or a small pocket sewn into each corner for hanging the antenna, as well as a pocket with a hook-and-loop fastener over the insulator. The pocket will allow access for tuning the antenna by adjusting the hairpin spacing.

## Usage

A glove antenna provides a practical solution for unique operating conditions. A portable antenna like this one does not need to be placed high up in the air. If you are backpacking, being able to quickly hang or take down an antenna from a tree is a benefit. In windy weather, the glove may blow around, with a minor effect on performance. Regardless, it is still much better than the stock antenna that comes with a handheld. There is some loss in the steel elements that may cost a dB or two in efficiency, but make the match broader. I have the antenna well-matched in free space at 146 MHz. When near other metal objects or on a wall, the resonant point drops, though my radios can accept the SWR change in most cases. For use in emergencies, go-kits, or local shelters, hanging the glove on an interior wall can make the difference when trying to hit a local repeater. You will likely find the compromises for this antenna to be acceptable when compared to the benefits. I thank Gary Bishop, NQØV, for helping me with the initial testing.

---

Jim Grebey, N4KCT, is the author of *Red Tide: Life on the Martian Frontier*. He became active in amateur radio in 1964 as WN2YTF. A year later, he upgraded to General class and became WB2YTF before earning his Advanced-class license in 1967. In 2021, he relicensed to General class, upgrading to Amateur Extra class the following year. He can be reached at **jim@grebey.com**.

# CHAPTER 4

# Low Profile Wire Antennas

# A Low Profile Two-Element 40 Meter Wire Beam

*Try this fixed direction wire beam on Field Day.*

Larry Serra, N6NC

Antennas and tower heights are restricted where I live. My flat roof is 35 feet above ground level (AGL), and encloses an approximately 38 × 68 foot rectangle, with a 30 inch high parapet wall. The roof already contains my SteppIR™ BigIR vertical for 80 and 40 meters. All my roof-mounted antennas are "temporary" so they can be broken down and removed if necessary.

The long dimension of the roof rectangle runs Northwest-Southeast, so an antenna lobe pattern perpendicular to that would aim Northeast. John Barcroft, K6AM, persuaded me to try a simple two-element 40 meter wire Yagi antenna oriented along the long

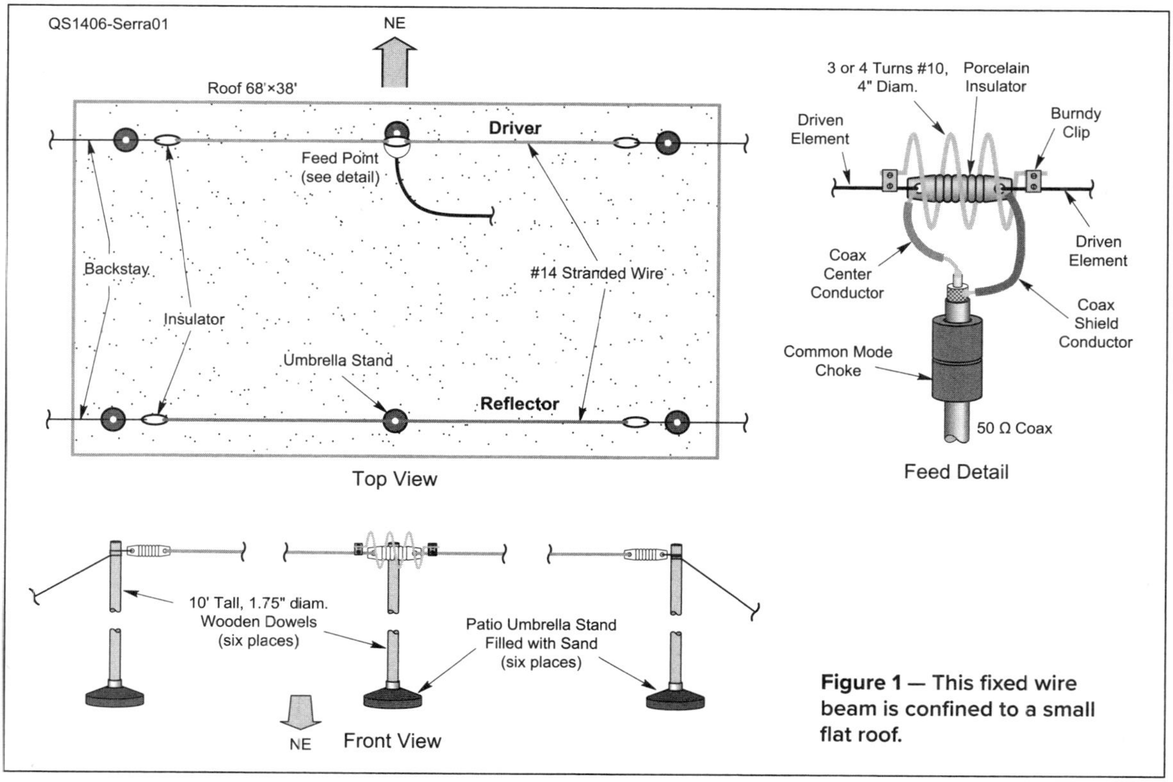


**Figure 1 — This fixed wire beam is confined to a small flat roof.**

dimension of the roof. Because I live in San Diego, California, a broad beam width antenna aimed NE is suitable for domestic contests across North America, as well as providing good coverage to much of Europe.

## Building the Wire Beam

With help from Ed Ross, N6GZI, I fashioned a 40 meter dipole driver element from #14 AWG stranded copper antenna wire and resonated it around 7.038 MHz, favoring the CW end of the band. To mount this antenna as unobtrusively as possible, I obtained six 10-foot, 1.75-inch diameter wooden dowels, and painted them with two coats of flat (not glossy) Rust-Oleum® for weather protection. I also purchased six free-standing umbrella base stands and filled the bases of the stands with about 52 lbs of sand or water.[1] That's enough weight to hold the wire antenna in place for my local wind conditions. I painted the dowels and drilled them each with a ¼-inch hole 2 inches from the top, then used short PVC pipe sections to bring the bottom dowel diameter up to the diameter of the patio umbrella stand pipe for a snug fit.

## The Driver and Reflector

We used a center insulator for the driven element, but you can use a dipole center connector. We mounted the dipole driver with porcelain end insulators on three evenly spaced dowel masts along the northeast side — the long dimension of the flat roof (see **Figure 1**). Use Burndy clips (two conductor wire connectors) on each wire end so you can adjust the wires to resonant length with an antenna analyzer before soldering the wires in place.[2] I installed Burndy clips on the wire legs 3 to 4 inches on each side of the driver dipole element center connector to connect a matching coil. I wound the four-turn coil with U/V resistant insulated #10 AWG solid copper house wire around a 4-inch diameter form, leaving leads long enough to reach the Burndy clips, then placed the coil around the center dipole insulator between the Burndy clips. My *EZNEC* model showed driven element legs of 33.25 feet each, and a reflector element length of 69.83 feet. I couldn't quite fit those lengths because of physical constraints on the roof size, so both element lengths needed field adjusting. Of course, the exact dimensions for your specific conditions may vary.

Roof obstructions forced us to space the elements no more than about 37 feet apart. I used one continuous length of #14 AWG antenna wire for the reflector, and threaded the wire through the hole in the center dowel then secured it to each end dowel with a small porcelain insulator. It's a good idea to use UV-resistant poly cord as backstays at each end to tension the driver and reflector wires so they are straight and flat.

## Matching the Antenna

We checked the VSWR at the driven element. From these measurements, John, K6AM, estimated that for the purpose of matching, the *modeled effective height* using *EZNEC* should be 15 feet AGL. That would approximately account for all the metal and antennas on the roof, and for the three-story building between the antenna and actual ground. Separately, the antenna should be modeled at the actual height of 45 feet AGL to estimate antenna patterns.

To minimize the VSWR, spread or compress the matching coil. Ideally we should have readjusted the resonant length of the driver dipole after we added the matching coil (turns), but we arrived close to the proper length by the cut-and-try method. I installed a common mode choke on the coax at the antenna feed point. [See "Choke Baluns" in *The ARRL Antenna Book*.[3] — *Ed.*]

## Results

The *EZNEC* model of the 40 meter wire beam predicts a peak gain of 9 to 10 dBi at 50 degrees elevation angle. See Ward Silver's, NØAX, *Antenna Modeling For Beginners* for more details about antenna modeling.[4] We estimate that adding the reflector to a dipole increases the antenna gain about 3 dB over a dipole at the same height. The low antenna height pushes the maximum gain lobe up to an elevation angle of 50 degrees. The horizontal beam width is about 100 degrees.

In practice, signals received on the wire beam are consistently stronger with my Elecraft K3 transceiver than on my vertical antenna. We compared my vertical and wire beam using my *CW Skimmer* SDR and the Reverse Beacon Network through the *Skimmer* Aggregator.[5] My 40 meter wire beam consistently outperformed my vertical antenna all across the US, Russia, Ukraine, and continental Europe. The broad beam width (100 degrees) enables me to copy signals in the sector from Alaska through Venezuela from my NE beam orientation in San Diego, CA. I also receive signals from the Pacific, China, and Japan from the back lobe of the beam.

## Conclusion

You can erect this stealthy antenna using lighter-strength materials to present an even lower profile. The design would also work on large decks and docks, and could be used suspended from trees, sports field light standards, and flagpoles, and in many portable locations for Field Day. You can expect good performance with this simple wire antenna.

---

**Notes**

[1]Patio umbrella stand — product 11527452, **walmart.com**.

[2]Burndy clips — **homedepot.com/p/Blackburn-Type-ADR-2-Conductor-1-Hole-Mount-Wire-Connectors-for-1-0-Stranded-Max-Wire-2-Pack-ADR11-21-B2-5/100172941**.

[3]ARRL order no. 6948, available from your ARRL dealer, or from the ARRL Store. Telephone toll-free in the US 888-277-5289, or 860-594-0355, fax 860-594-0303; **arrl.org/shop/**; **pubsales@arrl.org**.

[4]ARRL order no. 3961.

[5]**reversebeacon.net**

---

Amateur Extra class license holder Larry Serra, N6NC, has been licensed for 55 years. He has won numerous international, W6, and San Diego Section CW contests. He was a member of the 6E2T contest team that won the World M/2 class in the 1995 ARRL DX CW contest from Ensenada, Mexico. Larry is a lawyer specializing in commercial law, land use, and insurance coverage. He is law professor and author of four books about his experience as a naval intelligence officer in Cambodia during the Vietnam war (**nilohatien.com**). He has also held a USCG Masters license, and has sailed the Atlantic, western and eastern Pacific Oceans, and the Caribbean Sea in the US Navy, as a US Merchant Mariner, and as a cruising yachtsman. Larry can be reached at **hlserra@gmail.com**.

# A Stealthy Quad Vertical Antenna

*This antenna has quarter-wave elements on the 40 – 10 meter bands, and is deer-friendly.*

Stan Ekiert, K3KKH

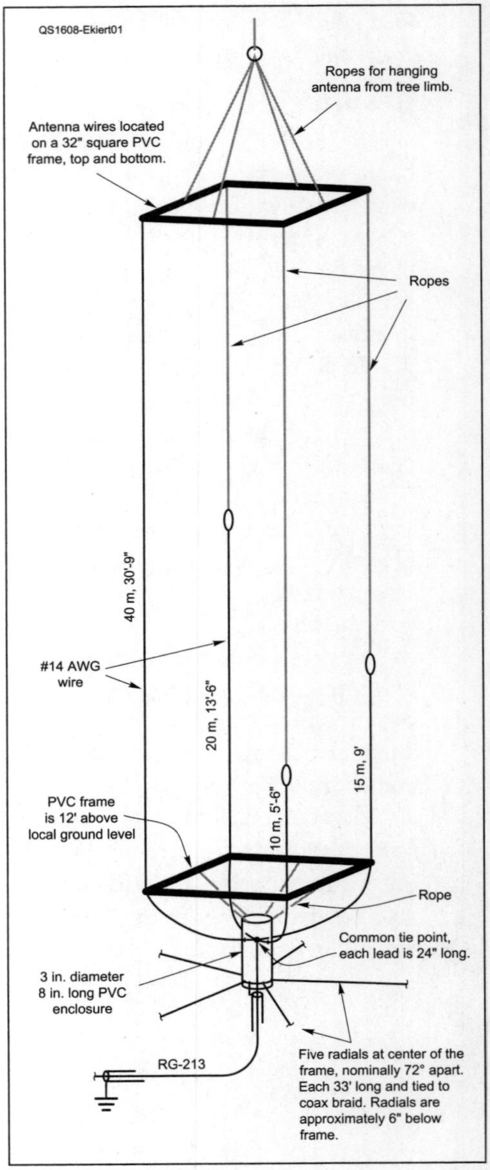


**Figure 1 — The Quad Vertical Array has four quarter-wavelength long-wire verticals cut for the 40, 20, 15, and 10 meter bands. The verticals are supported by two PVC frames that are 30 feet, 9 inches apart.**

Returning to the airwaves after a 50-year hiatus was a bit daunting due to having to learn about all the new equipment, new bands, and new modes. What provoked my anxiety the most were the restrictions on antennas within our townhouse community. Ham radio antennas are not allowed. Determined to get back on the air, I considered both indoor and outdoor antennas. I discounted an attic antenna because our attic contains a furnace with heating ducts that would affect any close-by antenna. The only reasonable approach was to use the very dense woods behind our house to provide the necessary camouflage for an antenna. I began looking at configurations that could be adapted to the dense woods, and that met my set of requirements.

My four simple requirements for the antenna were, (1) an outside antenna must be stealthy, (2) the antenna must function on the traditional bands of 40 through 10 meters, (3) no external band switching, and (4) the antenna must be high above the ground to avoid deer traffic. I also wanted to build the antenna myself.

## The Antenna

I considered the fan dipole, but I wasn't sure that I could string the dipole through the dense

Figure 2 — The PVC enclosure can be seen centered below the lower PVC frame.

trees without attracting too much attention. I thought that the fan dipole approach could be applied to a quarter-wave vertical, which would be easier to hang from a tree branch. I settled on four closely spaced quarter-wave verticals, all simultaneously fed with RG-213 coaxial cable. I named the configuration a Quad Vertical Array (QVA, see **Figure 1**). It consists of four subassemblies: an upper and lower frame, four wire verticals, five wire radials, and a small enclosure that houses the electrical connections between the verticals, radials, and the coax feed line. The vertical wires are located at the corners of the 32-inch square frames.

## Frame Construction

I constructed the upper and lower frames each from four 30-inch-long sections of half-inch diameter PVC pipe. I reinforced the pipes with pine dowels to stiffen the assemblies, and glued the pipes into 90-degree elbows to complete the square frames. The frame size was dictated strictly by convenience rather than a formula. I purchased two 10-foot lengths of PVC pipes from Home Depot. They were cut in half by Home Depot into 5-foot sections that fit in my car. At home, I cut each section in half again for the final length of 30 inches. With the PVC elbows at each corner, the final dimension was 32 inches on an edge. For each frame, I drilled a ¼-inch hole in each elbow and fitted a 2.5-inch-long ¼-20 eye bolt, and used 1.5-inch

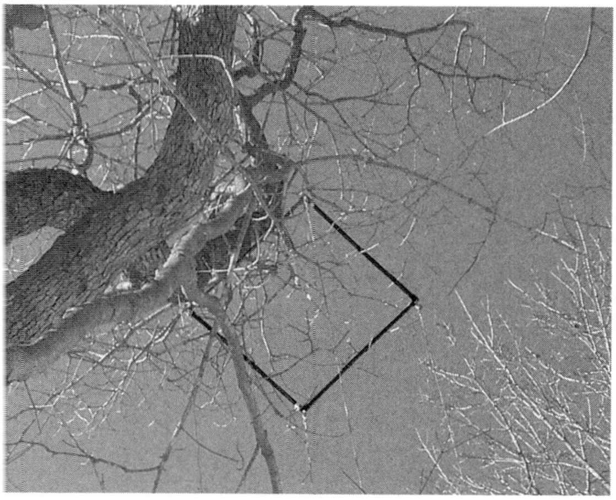

Figure 3 — The upper frame hangs 43 feet above the ground. It is supported by monofilament, 300-pound test strength fishing line.

Figure 4 — The 3-inch-diameter PVC enclosure is supported by rope tied to eye bolts on the lower frame and on the enclosure. The wire radials were looped through the enclosure eye bolts several times for strain relief before entering the enclosure through holes adjacent to the bolts.

**Table 1**

**Initial design values for the QVA verticals favor the lower portion of each band. Lengths include the 24-inch jumpers below the lower frame.**

| *Frequency, MHz* | *Length* |
|---|---|
| 7.15 | 32 ft, 9 in |
| 14.1875 | 16 ft, 6 in |
| 21.2375 | 11 ft |
| 28.5 | 8 ft, 3 in |

quick-links to attach the verticals to the eye bolts via cable thimbles at each end of the wire verticals, as seen in **Figure 2**.

Four additional ¼-inch holes at the midpoint of each pipe in the upper frame were fitted with ¼-20 eye bolts to provide a rope connection between the frame and a 2-inch iron ring from which I hung the entire antenna structure from a tree branch 53 feet above the ground.

## Antenna Construction

I initially cut the lengths of the #14 AWG copper wire verticals for 40, 20, 15, and 10 meters based on the formula

$L(\text{ft}) = 234/f(\text{MHz})$ [Eq 1]

**Table 1** shows the initial design frequencies and lengths for each band. Later, during the tuning process, I changed some of the frequencies and corresponding lengths. The distance between the top and bottom frames is 30 feet, 9 inches and is determined by the 32-foot, 9-inch length of the 40 meter vertical, which includes 24 inches of wire needed to connect the bottom of each vertical to a common mounting plate within the small PVC enclosure. To maintain the frame separation, the remaining verticals consist of a combination of antenna wire plus a length of rope (see **Figure 3**).

I cut each vertical to the length shown in Table 1, plus an additional 5 inches to allow for wrapping the wire over two cable thimbles and twisting the wire around itself to secure it.

## The PVC Enclosure

The small PVC enclosure consists of an 8-inch-long, 3-inch-diameter PVC pipe with two end caps (see Figure 2). The end caps are secured to the pipe by four ¼-20 thumb screws with external nuts. The external nuts were tightened against the outside of the end caps to prevent the thumb screws from loosening. The thumb screws allow easy access to the interior of the enclosure to address changes such as adding additional radials. In the final assembly, I sealed the end caps with silicone caulk. The enclosure houses two 2.5-inch square aluminum plates with a corner of each plate bent 90 degrees so they could be attached with screws to the interior of the enclosure. The upper plate functions as the common attachment for the vertical wires and the center wire of the feed line coax (see **Figure 4**).

## Radials

The lower plate is the common attachment for five 33-foot radials, and the coax braid. All connections to the plates use 8-32 screws, washers, lock washers, and nuts. Five 2.5-inch-long ¼-20 eye bolts were mounted at 72-degree intervals around the periphery of the enclosure to provide strain relief attachments for the radials. To support the enclosure, sections of rope were tied between the eye bolts on the lower frame and four of the five eye bolts on the enclosure. I mounted an SO-239 connector on the lower end cap to provide the connection to the RG-213 feed line from the shack.

The ends of the radials are attached to ropes that pass through screw eyes that are screwed into distant trees — 10 to 12 feet above the ground to avoid deer traffic. The five radials are nominally distributed throughout the woods at increments of 72 degrees about the axis of the QVA. The end of each rope is attached to a 1

**Table 2**

**Tuning sequence of the 20 meter vertical based on Eq 3.**

| *Iteration (Fig. 5)* | *SWR min, MHz* | *dL, ft* | *df, MHz* | *df, meas.* | *New min. SWR, MHz* |
|---|---|---|---|---|---|
| #1 | 13.5 | –0.25 | 0.195 | 0.2 | 13.7 |
| #2 | 13.7 | –0.25 | 0.201 | 0.2 | 13.9 |
| #3 | 13.9 | –0.25 | 0.206 | 0.1 | 14.0 |
| #4 | 14.0 | –0.25 | 0.209 | 0.2 | 14.2 |
| #5 | 14.2 | – | – | – | – |

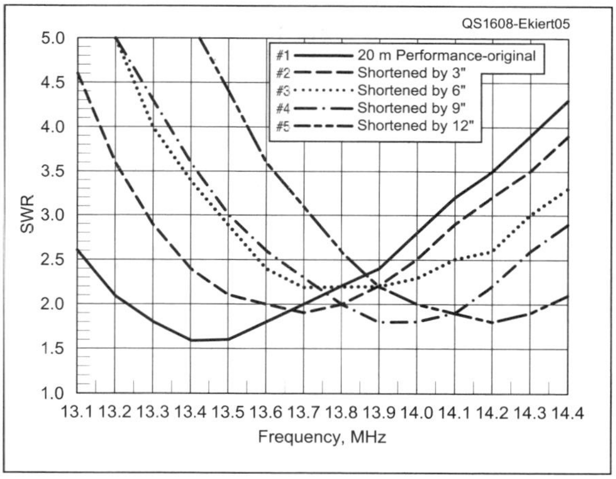


**Figure 5 — SWR plots for four cycles of tuning the 20 meter antenna; #1 is the initial antenna, #2, #3, #4, and #5 show the SWR of the four cycles, with #5 being the final result.**

liter bottle filled with sand, which provides 4 pounds of tension to keep each radial taut. The radial tensioners also accommodate tree movement due to wind without adding strain on the radials or the lower frame.

## SWR Performance

I measured the SWR of each vertical with an MFJ-259B antenna analyzer through a 10-foot length of RG-213 cable. Initially, the best SWR on the 20 and 10 meter antennas was well below each band. The 20 meter vertical displayed a minimum SWR of 1.6 near 13.5 MHz as seen in **Figure 5**, while the 10 meter vertical displayed a minimum SWR of 1.3 at 26.4 MHz.

### Tuning the Verticals

Prior to making any adjustments to the 20 and 10 meter antennas, I wanted to know how the frequency of the SWR minimum would change for a given change in length. Knowing what to expect would minimize my tuning effort, because tuning these verticals is a two-step process. The wire is first adjusted, then the rope is adjusted to preserve the frame separation. For each cycle I lowered, tuned, then raised the QVA for a new SWR measurement. To estimate the change in length for a change in frequency of the SWR minimum, I differentiated Eq 1 with respect to frequency,

$$dL/df = -234/f^2 \quad \text{[Eq 2]}$$

*dL* is the change in antenna length (feet) that results from a *df* change in frequency (MHz), so

$$dL = -234df/f^2 \quad \text{[Eq 3]}$$

Solving for *df* gives the change in frequency (MHz) when the antenna length is changed by *dL* feet.

$$df = -f^2dL/234 \quad \text{[Eq 4]}$$

*f* in MHz is the frequency of the SWR minimum that needs to be moved, and is selected from a plot of the SWR versus frequency (see Figure 5).

When *shortening* an antenna wire length, select the *highest* frequency for which the SWR is a minimum. When *lengthening* an antenna wire length, select the *lowest* frequency for which the SWR is a minimum. Tuning should start with the lowest frequency antenna and work towards the highest frequency antenna in order to minimize harmonic interaction. **Table 2** shows the results of four tuning cycles for which the length was reduced in

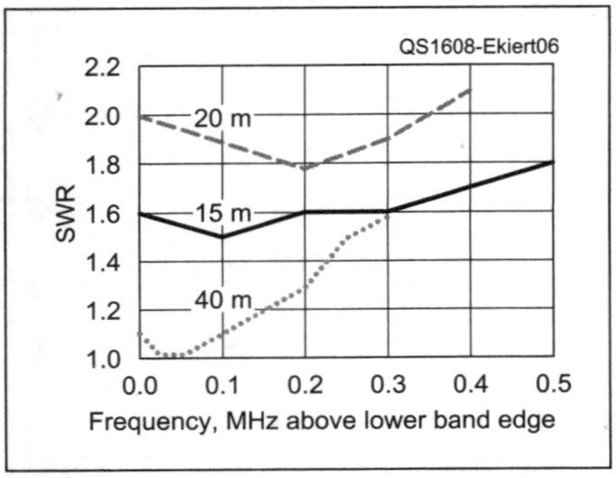


**Figure 6 — Final SWR plot for the 15, 20, and 40 meter antennas.**

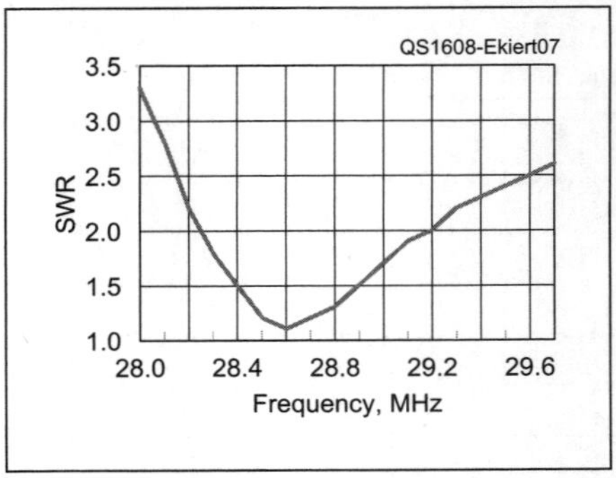


**Figure 7 — Final SWR plot for the 10 meter antenna.**

3-inch intervals (*dL* = –0.25 feet) to arrive at the desired SWR minimum frequency of 14.2 MHz. Figure 5 shows the SWR plotted for each cycle.

The second column in Table 2 shows the frequency of the SWR minimum that must be moved to a higher frequency. The *dL* column shows the change in feet of the antenna length. The *df* column predicts the frequency change of the SWR minimum due to the length change *dL*. The last two columns show the actual measured frequency change, and the frequency of the new SWR minimum, respectively. Table 2 shows that Eq 3 does a good job of predicting the new minimum SWR frequency.

**Figure 6** shows the final SWR in the 15, 20, and 40 meter bands, and **Figure 7** shows the SWR in the 10 meter band. **Table 3** shows the final antenna lengths and frequencies of minimum SWR.

**Table 3**

**The final values for the QVA verticals. The first column is the frequency of the SWR minimum.**

| *Frequency, MHz* | *Length* |
|---|---|
| 7.05 | 32 ft, 9 in |
| 14.2 | 15 ft, 6 in |
| 21.1 | 11 ft |
| 28.6 | 7 ft, 6 in |

## Summary and Final Thoughts

I found the solid #14 AWG wire very stiff and difficult to manipulate during construction, especially when trying to mount the two aluminum plates to the inside wall of the small PVC enclosure. The upper plate has five wires attached to it, while the lower plate has six. Both assemblies were extremely rigid and difficult to align with their mounting holes in the enclosure. Stranded wire might be a better choice. In retrospect, I would also add a few radials resonant on the 20 and 10 meter bands. This should help to lower the SWR response and increase bandwidth.

My Quad Vertical Array has been operational since September 2014 and has exceeded my expectations. I have used it with an antenna tuner on all ham bands from 40 through 10 meters. It is very stealthy in all seasons, with no complaints from our deer population.

Photos by the author.

Stan Ekiert, K3KKH, was first licensed as a Novice in 1960 as KN3KKH, and later earned the General class license K3KKH. In 1961, he earned a Commercial First Class Radio Telephone Operators License. Stan has a BSEE and MSEE from the University of Pittsburgh. He worked for Eastman Kodak Company for over30 years, on government programs involving imaging sensors and systems, and demonstrated the first solid-state imaging using photodiode arrays in 1970. In 1998 he joined ITT Corp — now Harris Corp — in Clifton, New Jersey, where he worked on SINCGARS radio enhancements, GPS navigation payloads, and electronic warfare programs until retiring at the end of 2012. After a 50-year hiatus, he returned to the airwaves in 2014 and is thoroughly enjoying operating on 80 through 10 meters. Stan can be reached at 35 Spring Hill Cir, Wayne, NJ 07470, and **roadspyder@msn.com**.

# Inexpensive Indoor HF Antenna for Apartments

*A horizontal loop strung around a room can get you on multiple ham bands.*

Al Sweet, KG6HM

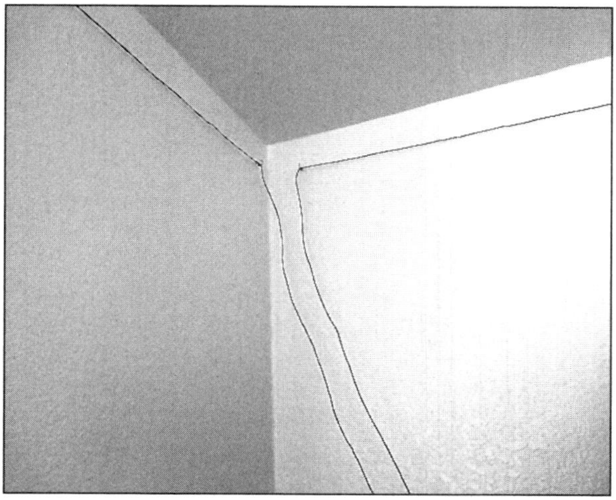

I live on the third floor in an apartment building that does not allow outdoor antennas. To satisfy my desire to operate the *WSJT-X* modes, I needed an inexpensive, nondirectional antenna that worked well indoors and did not need an Earth ground. After some experimenting, I tried a horizontal loop antenna strung around the ceiling molding of my apartment bedroom. Such an antenna takes up no floor space and can be fabricated with simple hookup wire. The loop antenna requires a balanced feed, which eliminates the need for any kind of ground. An antenna tuner is required to adjust the antenna for each band of operation.

## Building and Calibrating the Loop

The size of the loop corresponds to the size of the room and the bands I most wanted to operate. My apartment bedroom is approximately square with a length and width of about 15 feet (5 meters), or about 60 feet (20 meters) of total antenna periphery, which is close to resonance in the 20-meter band. The loop antenna is held in place with picture hooks at the ceiling molding.

I operate this antenna on the 80- through 6-meter amateur bands via an SWR meter and an MFJ-901B Versa Tuner (see **Figure 1**) by adjusting the tuner for best match on each of my operating bands. The tuning procedure only needs to be done once. This MFJ tuner has three different adjustments — a "transmitter" adjustment setting, an "inductance" adjustment, and an "antenna" adjustment setting. I recorded my tuner settings in **Table 1**, and posted the settings next to the radio. When changing bands, I simply reset the tuner to the adjustments associated with the desired band. You might want to fine-tune the settings a bit to get the minimum SWR at your frequency of interest. **Figure 2** shows example SWR traces on the 80- and 20-meter bands on the transmitter side of the antenna tuner.

## Operating Power Levels

I tested my horizontal loop at power levels up to 100 W on SSB and CW. [See the sidebar "RF Exposure Considerations." — *Ed.*] The

**Table 1**

**The MFJ-901B Versa Tuner control settings for my loop antenna.**

| *Frequency (MHz)* | *"Antenna" setting* | *"Inductance" setting* | *"Transmitter" setting* |
|---|---|---|---|
| 3.576 | 4.2 | G | 2.5 |
| 7.076 | 1 | G | 5.6 |
| 10.139 | 3 | E | 6 |
| 14.076 | 1 | A | 1.9 |
| 18.102 | 1 | C | 6 |
| 24.917 | 2.5 | A | 1 |
| 28.076 | 5 | B | 5 |
| 50.276 | 5 | A | 1 |

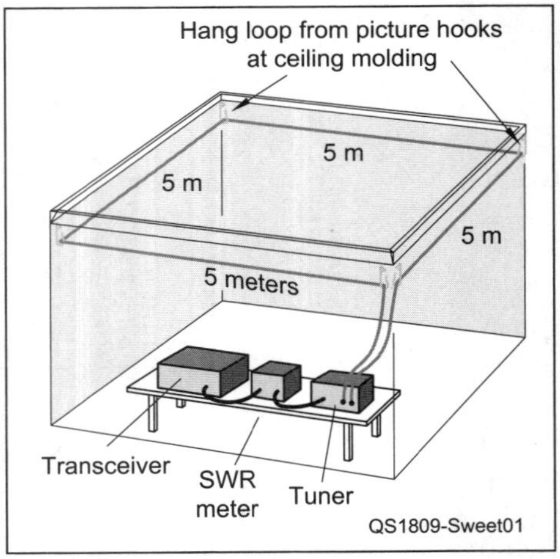


**Figure 1 — This horizontal loop antenna is hung on picture hooks near the ceiling molding of my 5 × 5 meter room. The transceiver connects through the SWR meter to a tuner that has a balanced output.**

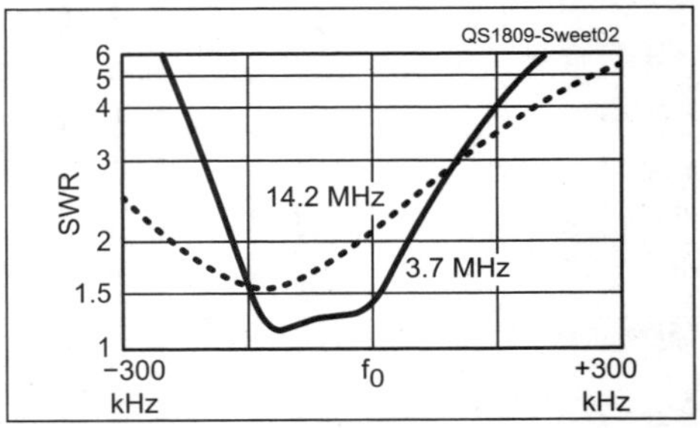


**Figure 2 — Sample SWR traces show the match on the 80 – and 20-meter bands.**

## RF Exposure Considerations

**Operating close to an indoor antenna requires a cautious approach.**

By filling out and signing Form 605 for license application, modification, or renewal, all Amateur Radio operators certify that they *have read* and *will comply* with Section 97.13(c) of the Federal Communication Commission's Rules (available at **gpo.gov/fdsys/pkg/CFR-2011-title47-vol5/pdf/CFR-2011-title47-vol5-sec97-13.pdf**) regarding Radiofrequency (RF) Radiation Safety and the Amateur Service section of OST/OET Bulletin Number 65, available at **fcc.gov/oet/info/documents/bulletins/**.

Unusual antennas like this indoor loop can be evaluated using an NEC model. Be sure to evaluate the total fields inside the room, as well as in regions external to the room, like on the other sides of the walls. — *Ed., on behalf of the ARRL RF Safety Committee*.

results were very gratifying, with great signal reports from many parts of the country and the world. This means that the same indoor horizontal loop antenna can be used effectively at low power for FT8, JT65, and CW, and at higher power SSB and CW.

Several members of my radio club (Amateur Radio Club of Alameda in California) have constructed similar antennas in their homes and are evaluating the performance. I hope other apartment dwellers will try this simple and effective low-cost HF indoor antenna and let me know how it works for them.


Al Sweet, KG6HM, built his first crystal radio set in 1954 at age 11, and received his Novice license, KN1DKF, in 1957. He operated a 15 W homemade tube transmitter, getting hooked on DXing for life. Al now holds an Amateur Extra-class license. He graduated from Worcester Polytechnic Institute with a BS in electrical engineering, and later received MS and PhD degrees in electrical engineering and physics from Cornell University. Al has operated a consulting service firm devoted to the design of RF integrated circuits. He is also an adjunct professor of electrical engineering at Santa Clara University, and is the principle investigator of a nanotechnology research project at the Lawrence Berkeley National Laboratories Molecular Foundry Nanotechnology center in Berkeley, California. You can reach Al at **allensweet@aasweetphd.com**.

# A Coaxial Cable Vertical Dipole Antenna

*Hang it from a tree limb and you're on the air!*

Al Yerger, K2ATY

Vertical antennas have long been utilized for communications throughout the spectrum for any number of reasons. An ideal vertical antenna has a low angle of radiation, keeping most of the transmitted energy close to the horizon, providing excellent long distance propagation. In addition, they provide omnidirectional radiation patterns and, depending upon the frequency and construction, can be useful in limited space situations and for portable operating.

Vertical antennas also have some disadvantages. Among them is the requirement for a rather robust ground system when the antenna is ground mounted, or a system of elevated ¼ wavelength radials if the antenna is mounted above the ground. Another problem with a traditional ¼ wavelength vertical is that most of the radiation from a vertical antenna takes place near the bottom of the antenna where the antenna current is highest. Unless the antenna is mounted high and in the clear with the requisite radials, most of your signal is subjected to various forms of ground clutter, such as nearby buildings. Also if your ground plane is not perfect, the maximum radiation will not be at the low angle you want and expect. In an attempt to raise the point of maximum radiation and solve some of these problems, end-fed ½ wavelength verticals are sometimes used, but they present high feed point impedances along with other matching issues.

Another approach to a vertical ½ wavelength antenna is the coaxial vertical. A coaxial vertical is simply a ½ wavelength dipole (**Figure 1A**) oriented vertically. The lower ¼ wavelength portion of the antenna is hollow and the transmission line is run inside the radiator. This design has been used for years in VHF and UHF antenna design. Figure 1B shows the construction of a basic ½ wavelength coaxial vertical.

For HF antennas this design is not always practical due to the size of the antenna. For bands below 10 meters, a coaxial vertical of conventional design may be simply too heavy to properly support. The design we are presenting here combines the properties of a line isolator, acting as a common-mode choke, along with some basic transmission line theory to create a practical HF coaxial vertical antenna. The lower frequency limit of the design will be dictated by the height of the available support.

## Skin Effect

To start the discussion regarding the theory behind this antenna, we need to ask a question that I am sure everyone has heard some-

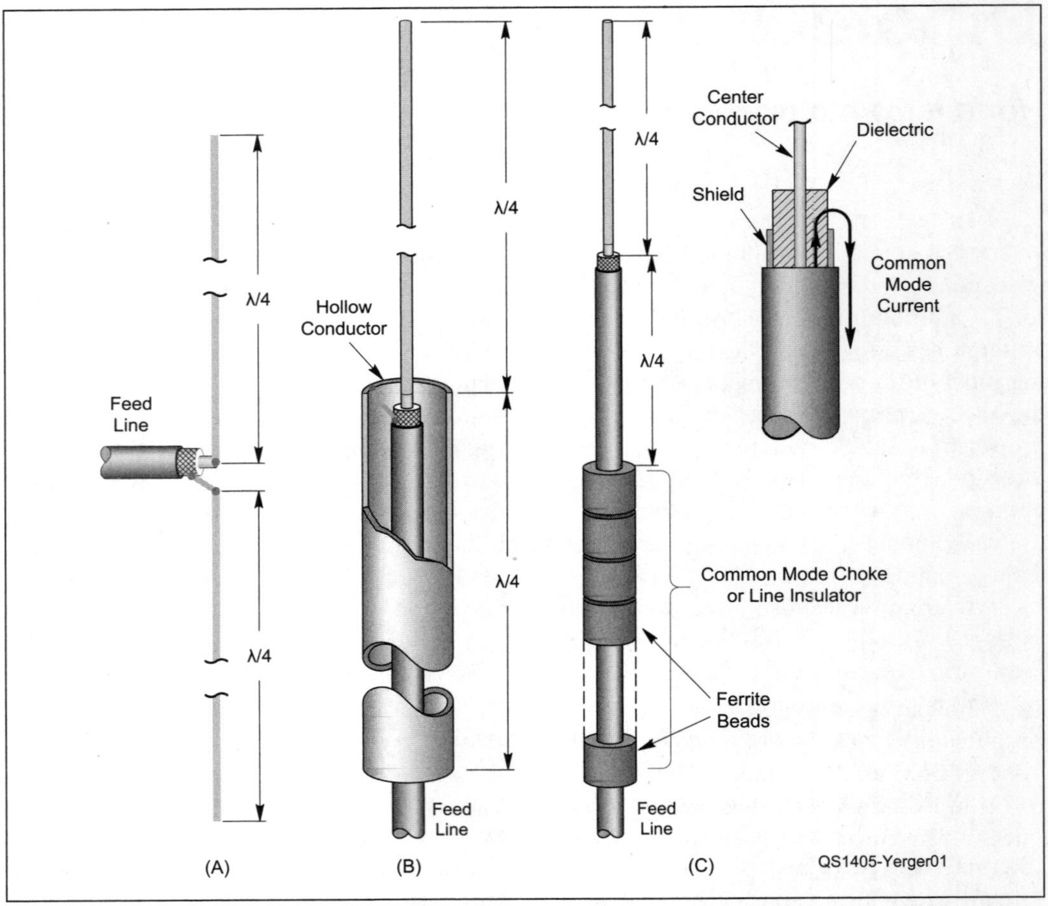


**Figure 1** — Evolution of the coaxial cable vertical dipole: At (A), the vertical half-wave dipole. Because it is a dipole, a ground radial network is not required, but care must be taken to keep the feed perpendicular to the dipole to prevent coupling onto the feed line. At (B), the coaxial vertical dipole. "Coaxial" refers to the tubular bottom element, which serves to route the feed line to the center of the dipole, which avoids feed line coupling. At (C) the coaxial cable vertical dipole that uses a portion of the feed line shield to serve as the bottom element of a vertical dipole. The feed line shield is unconnected at the center of the dipole so current flows from the shield's inside to its outside, where it flows down as a common mode current until it encounters a high impedance presented by a common mode choke, acting as a line isolator, place a ¼ wavelength away.

| Table 1 | | |
|---|---|---|
| **Element Lengthsl** | | |
| *Frequency, MHz* | *Lower Element (ft in)* | *Upper Element (ft in)* |
| 3.750 | 58' 9" | 62' 3" |
| 7.150 | 30' 6" | 32' 6" |
| 10.125 | 21' 5" | 22' 11" |
| 14.175 | 15' 2" | 16' 4" |
| 18.115 | 11' 10" | 12' 10" |
| 21.225 | 10' 1" | 11' 0" |
| 24.950 | 8' 7" | 9' 4" |
| 28.300 | 7' 6" | 8' 3" |

where; does current flow *on* or *in* the shield of the coax? The answer to that age-old question is yes and no. The reason for the ambiguous answer lies in the operation of coaxial transmission line and a phenomenon known as *skin effect.*

Skin effect is the principle that says RF energy tends to flow on the surface of a conductor. The amount the energy penetrates into the conductor depends upon frequency. The higher the frequency, the less the energy penetrates into the conductor. From a practical standpoint, we can assume that all of the RF energy flows on the surface.

Now let's take a look at coaxial cable. In the upper right corner of Figure 1C, you'll see the construction of typical RG-213 type coaxial cable. There is a dielectric and then an inner conductor surrounded by an outer conductor or shield. Applying the skin effect principle, we see that current flows on the outer surface of the inner conductor and on the inner surface of the outer conductor. If the cable is properly terminated at each end, there should be no current flowing on the outer surface of the shield. The primary purpose of a balun or line isolator is to ensure that RF current does not flow on the outer surface of the transmission line.

For this antenna, treat the inner and outer surfaces of the coaxial cable shield as if they were separate conductors. The outer surface of the shield becomes the sleeve portion of the coaxial vertical. To obtain resonance, the sleeve must end at the proper length to create a ½ wavelength dipole. To set the length of the sleeve, utilize a line isolator. I used a Radio Works T4 line isolator, but any high quality line isolator should work.

The line isolator places a very high inductive reactance in series with the outer surface of the cable creating essentially an open circuit. The location of this open circuit establishes the end of the lower element half of the coaxial vertical. To create the upper vertical element, we simply extended the center conductor with a ¼ wavelength piece of #14 AWG THHN copper wire from the local home center.

The current inside the coax flows up through the line isolator unaffected. When the current on the inside of the shield reaches the top of the coax it will then flow down the outside of the shield until it reaches the line isolator where, due to the inductance of the line isolator, the shield will appear as an open circuit effectively creating the lower end of the dipole. Due to skin effect, the two currents never "see" each other. This is how we cause the shield to function as if it were two separate conductors.

## Construction

See Figure 1C. To construct the lower half of the radiating element, start with a piece of RG-213 type cable. Cut the cable to the length shown in **Table 1** for the desired frequency. This length will be about 13% shorter than a ¼ wavelength in free space. This takes into account the fact that the diameter of the cable creates a somewhat thicker element than a single thin piece of wire. This lowers the resonant frequency of the antenna. The table values include a multiplier of 0.96 for the insulation on the cable. Also accounted for is the length of the connectors between the upper and lower elements. The top of the lower section really ends at the top of the PL-259 that connects the top section to the lower section of the antenna. This is where the current on the inside of the shield actually spills out, onto the outside of the shield. This transition adds about 1.875 inches to the overall length of the bottom section.

In other words, the lengths presented in Table 1 are the cut lengths of the cables before installing the connectors.

Prepare the lower section by cutting the proper length of RG-213 cable and installing a PL-259 connector on each end of the cable. Be careful that when you prepare the ends of the cable, you only expose the minimum amount of center conductor and shield required to properly attach the PL-259. Cutting away too much will raise the resonant frequency of the finished antenna. Adjusting the length of the lower element is difficult once the connectors are installed.

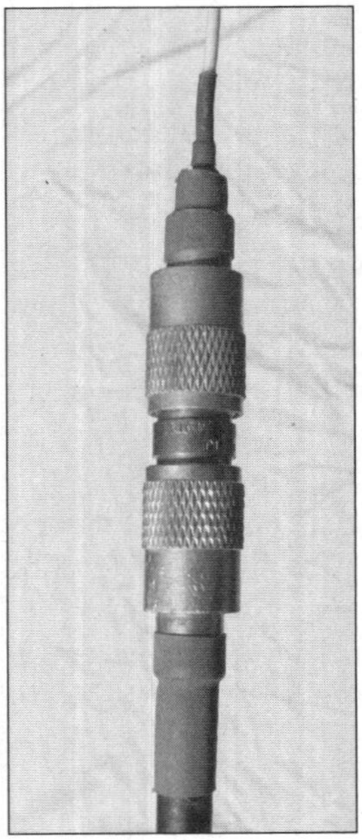

**Figure 2** — The top element, comprised of #14 AWG THHN house wire connected to a PL-259 plug that is joined to the bottom coaxial element through a PL-258 double female connector.

**Figure 3** — A typical installation of the coaxial cable vertical dipole hanging from a tree limb. The common mode current choke function is served by a Radio Works T-4 "line isolator."

A spreadsheet that will compute the parameters of the antenna is available at **arrl.org/qst-in-depth** in the **2014** dropdown.

For the upper element, measure the proper length of #14 AWG wire and solder it to the center pin of a PL-259 type UHF plug. Again, Table 1 contains suggested lengths for the various bands. I placed a small piece of the RG-213 dielectric around the bottom of the #14 AWG wire to help insulate the upper wire from the body of the PL-259. I then sealed the top of the connector with heat shrink tubing followed by an application of Scotch Kote®.

## Assembly and Tune Up

Connect one end of the lower cable assembly to the line isolator, join the upper and lower sections together with a barrel connector and then hang the antenna from a convenient support using a standard antenna end insulator. Feed the lower end of the line isolator with any convenient length of transmission line.

While the line isolator or common mode choke can provide a high impedance to shield current, it is important to recognize that the end of a half-wave antenna has a high electric field intensity that can couple around the choke and result in strange effects, including unexpected changes to tuning. Adding an additional choke some distance below the primary isolator can reduce this effect.

The antenna can be fine tuned to frequency by simply adjusting the length of the upper element. If you are careful when you construct the antenna, you shouldn't have to adjust the length very much to get a good match across the desired band. If you adjust the length of the upper element too much, you run the risk of unbalancing the antenna. Unless you get to an extreme condition, a little unbalance is harmless. The biggest effect you will probably notice is an increase in the SWR at the resonant frequency. As the feed point of a dipole antenna moves away from the center, the impedance of the antenna increases, increasing the SWR even though the antenna is actually resonant. The impedance of this antenna at resonance is probably not going to be exactly 50 Ω. The closest I could get to 50 Ω on my 17 meter prototype antenna was 69 Ω which is about 1.4:1 SWR. You may be able to obtain an exact match by experimenting with a coaxial matching section; however 1.4:1 still represents a decent match and is actually close to the theoretical impedance of a half wavelength dipole which is closer to 70 Ω than 50 Ω.

## Installing the Antenna

The antenna is ideally suited for portable operation. Just throw a rope over a suitable support, pull the antenna up and start operating. Because the antenna is a vertical radiator, it would be best to avoid hanging the antenna close to a tower or other metallic support. The antenna could also be suspended inside a fiberglass or PVC pipe.

## Variations

A possible variation of the antenna would be to build your own line isolator by placing about 12 inches of Amidon FB-77-1024 ferrite beads on a piece of RG-213 that would serve as the both the lower section and the transmission line (make sure that the beads stay together). Simply slide the beads to the proper position to set the end of the lower part of the coaxial dipole and install the proper top section. This would probably work better on the higher frequencies than it would down at 3.5 or 7 MHz, as it is harder to obtain the necessary inductance to choke off the RF current at the lower frequencies. Feel free to experiment and have fun.

---

Photos by the author.

ARRL member and Amateur Extra class licensee Al Yerger, K2ATY, was first licensed as WN2EHI in 1967 and held the call sign WA2EHI for 40 years. Previously with Motorola, Al is now an RF engineer with Bird Technologies Group specializing in interference mitigation and antenna site issues. Al is on the ARRL speakers list and is past Assistant Director of the Hudson Division. He has also served as president of the Crystal Radio Club and Director of Engineering for the Rockland Repeater Association. Al is interested in various communication modes and, with an EchoLink node at his station, he often operates via remote control over the Internet. You can contact Al at **ayerger@hvc.rr.com**.

# An Easy to Make Two Band HF Ground Plane

*Here's another application of the popular window line coupled resonator antenna.*

Joel Hallas, W1ZR

In the May 2011 issue of *QST*, we described a simple and efficient two band *coupled resonator* dipole made from window line. That antenna worked very well, was easy to make, and perhaps more importantly, easy to duplicate.[1] It became very popular and has since been adapted to most combinations of amateur bands from MF to UHF.[2] I recently received a request to further adapt it to vertical polarization.

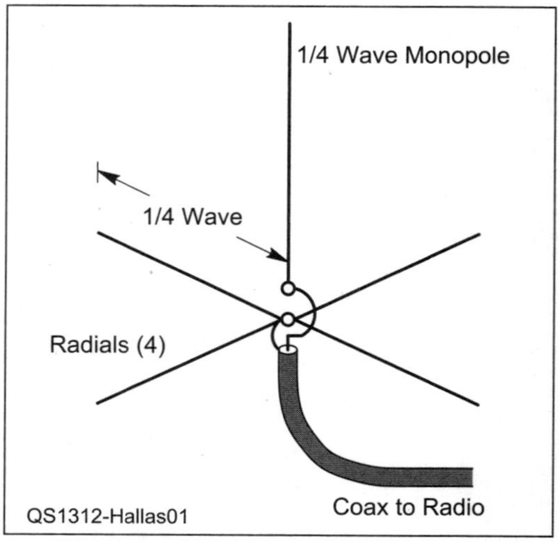


**Figure 1** — A traditional ground plane antenna. The four horizontal rods (radials) serve as an artificial ground for the monopole. Note that while they serve in place of a ground, they are not at ground potential. The ends of the radials are at the same potential as the tip of the monopole and people and animals should be protected accordingly.

## Shifting the Polarization

The original design could, of course, be adapted for use as a vertical dipole, and some have successfully done so, particularly for VHF FM operation. It turns out that a coupled resonator can also be configured as a vertical ground plane in order to minimize the height of the required support. This makes sense, since a vertical dipole needs a support more than half a wavelength high. With a single available support at that height, for most HF applications, an inverted V dipole provides improved performance.

### Try a Ground Plane

A *ground plane antenna* is a vertical ¼ wave monopole mounted and fed against an artificial ground system — the "ground plane." As

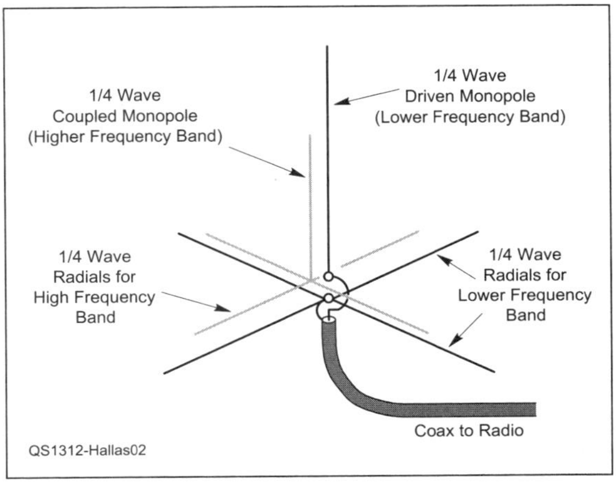


**Figure 2** — Simplified drawing of the two band coupled resonator ground plane. Note that there are no electrical connections to the higher frequency unit.

originally developed, the artificial ground consisted of a metallic disc with a radius of ¼ wavelength or more. While that is feasible at UHF, at lower frequencies, the ground plane is usually constructed of multiple ¼ wave conductors, called *radials,* perpendicular to the vertical element and located at its base, as shown in **Figure 1**. Configured in that way, the antenna has a feed point impedance that is half that of a dipole, or 36 Ω, in free space.

While the configuration of Figure 1 is often encountered — sometimes using a *gamma match* to provide a feed impedance of 50 Ω, another alternative is to lower the outboard end of the radials to make a *drooping* ground plane. This skeleton conical arrangement is topologically somewhere between the standard ground plane and a vertical conical dipole. Such an arrangement can provide a good match to 50 Ω coax. We have taken that approach here, although to use even less vertical space, if a small mismatch can be tolerated on the lower frequency band, horizontal radials can be used.

# The Coupled Resonator Ground Plane

As with the original two band horizontal dipole, this antenna is really two parallel ground planes made from window line. I selected the 20 and 15 meter bands for this model, however, it could easily be adapted to other pairs of bands. The lower frequency ground plane is fed directly with 50 Ω coax, while no portion of the higher frequency ground plane is connected directly to the driven unit, but receives its energy through parasitic coupling (see **Figure 2**). While three, four, or even more radials could be used, the two radial configuration I used is within less than 1 dB of being omnidirectional and works very well for those with physical constraints. Some have even used a single radial — picture a bent vertical dipole configuration — but that is more directional and has a higher angled main lobe, so we used two here.

## How High Should It Be?

As discussed in a recent *QST* article, a vertical monopole can be ground mounted, but works somewhat better if off the ground.[3] Another consideration is that the ends of the radials are a high voltage point, with a potential comparable to that at the top of the monopole. Thus for optimum safety, those ends should be high enough or otherwise protected from accidental contact. It is important to remember that while the radials serve as an artificial ground, in this configuration they are very much a part of the antenna and must be insulated as well as the vertical monopole.

If we were to have the ends of our radials at a safe height of 8 feet off the ground, and droop our radials at 45 degrees, we would need a support about 36 feet above ground. If we had a single support that high, we would

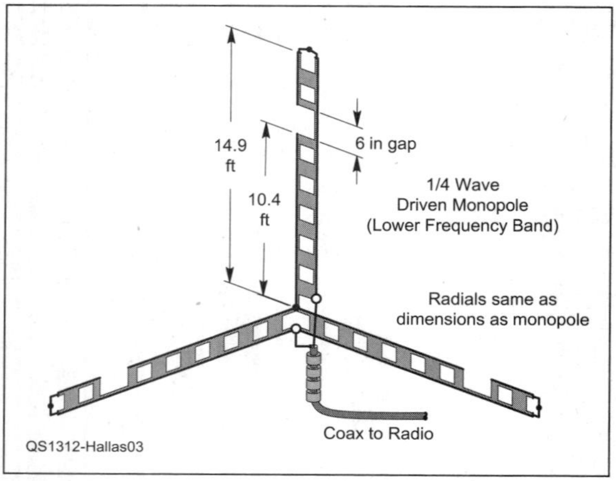


**Figure 3** — Dimensions of the window line 20 and 15 meter ground plane. For best match the radials should droop from the horizontal. While 45° is optimum, it will work satisfactorily with a less extreme droop.

**Figure 4** — Close up of the feed arrangement for the prototype ground plane. The six ferrite beads that make the common mode choke are shown below the feed point. A permanent antenna would do better with more structural integrity at the feed. The rope adjacent to the antenna is the halyard for a different antenna.

likely be better off with a dual band inverted **V** for these bands. Thus, in most installations a lower height will be used and sufficient insulation and other protection should be employed, especially if higher power is used.

## To Fold or Not to Fold

As with the coupled resonator dipole, it is possible to have either traditional linear elements or use the unused conductor of the window line beyond the end of the 15 meter elements to shorten the lower frequency antenna elements in a "folded" arrangement. At first, I was just going to describe the linear arrangement, but then decided that since available height was a key issue, I would save a bit more than a foot by using the folded approach. There really is no downside to this arrangement. Note that in the folded configuration, the highest voltage points are at the gap.

## Putting it Together

The exact dimensions will depend on the height above ground and the ground conditions. My dimensions (see **Figure 3**) were set with the base 12 feet above ground. I lowered the antenna so that the feed was at 3 feet, and the radial ends were at about 1.5 feet. As a result, the 20 meter resonance dropped by 270 kHz, while the 15 meter resonance dropped by only 20 kHz. Of course, this is a good direction in which to have a change, since it's easier to trim than stretch antenna wire. My suggestion would be to start with a few inches more than shown and trim as needed. By the way, order a few more feet of line than you think you will need — my order came up a few feet short.

I made my prototype from Davis RF window line that uses #18 AWG stranded copper covered steel wire.[4] While other types of window line could be used, different wire size or dielectric properties will have an effect on the final dimensions, so plan accordingly.

QS1312-Hallas05

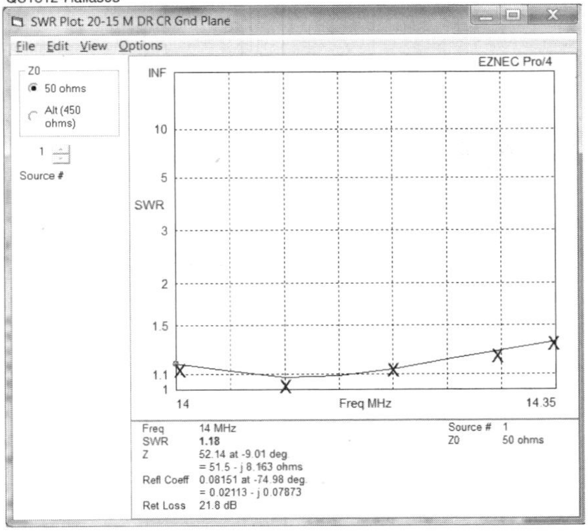


**Figure 5** — *EZNEC* predicted SWR of the folded coupled resonator ground plane on 20 meters. The measured results taken through 25 feet of RG-8X are shown with the Xs.

QS1312-Hallas06

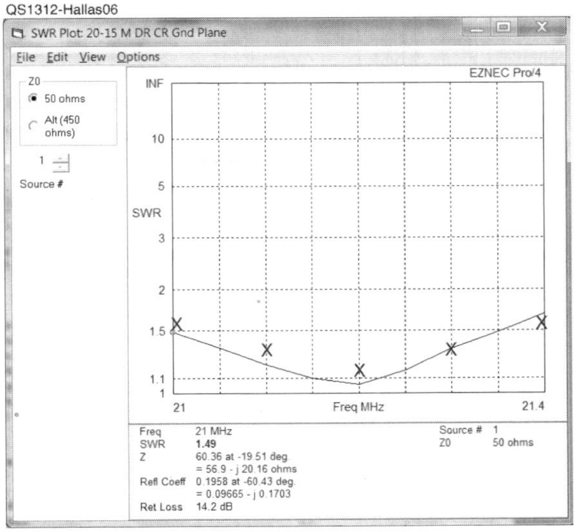


**Figure 6** — Same as Figure 5, but for 15 meters.

Note that while it looks like there are three distinct pieces, the two radials can be made from a single length of window line with the insulation stripped at the center of each piece.

I used a common mode choke made from six slip-on ferrite beads taped in place just below the feed.[5] Of course, other kinds of chokes could be used, however, the beads noted slip nicely over RG-8X, don't require tight turns as on a toroid and are lightweight and compact.

While the design shows the monopole and radials the same length, if you are close to being on frequency, you can just trim the monopole. It makes the antenna the ground plane equivalent of an *off-center fed* dipole. Any unbalance resulting from the small difference will be largely eliminated by the common mode choke (see **Figure 4**).

## Antenna Performance

The antenna came together quickly and works just as *EZNEC* predicted.[6] The measured SWR for both bands at the end of 25 feet of RG-8X is shown on top of the *EZNEC* SWR predictions in **Figures 5** and **6**. **Figures 7** and **8** show the elevation patterns on both bands, with the antenna feed 12 feet above typical ground.

Tests on the air tended to confirm the predictions. In A/B comparison with a three element triband Yagi at about 35 feet, the Yagi always did better by around 3 S units for stations at medium range that needed high elevation angles. DX stations were usually within 1 or 2 S units of the Yagi. Part of the difference was due to the extra 2 dB of line loss getting from my entrance panel to the antenna test position.

In summary, this antenna will not outperform a Yagi on a tower, but it will work DX at low cost and with a small footprint.

QS1312-Hallas07

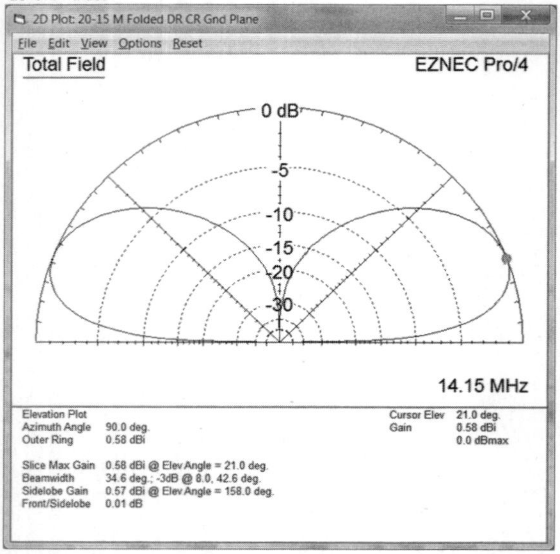


**Figure 7 — *EZNEC* predicted elevation pattern on 20 meters. The un-folded version (16.25 feet high) at the same height has a gain only 0.1 dB higher. The antenna was modeled with the base 12 feet above typical ground (conductivity 0.0005 S/m, dielectric constant 13). The azimuth pattern is within 1 dB of being omnidirectional.**

QS1312-Hallas08

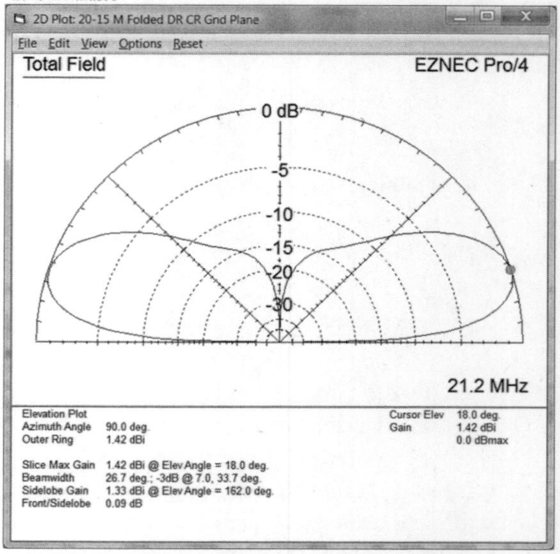


**Figure 8 — Same as Figure 7, but for 15 meters.**

## Notes

[1]J. Hallas, W1ZR, "Getting on the Air — A Folded Skeleton Sleeve Dipole for 40 and 20 Meters," *QST*, May 2011, pp 58-59.

[2]J. Hallas, W1ZR, "The Folded Skeleton Sleeve Dipole on Other Bands," *QST*, Oct 2011, p 48. An updated list is on the *QST* In Depth web page at **arrl.org/qst-in-depth**.

[3]J. Hallas, W1ZR, "How High Should Your HF Vertical Be?" *QST*, Nov 2011, pp 51-52.

[4]Davis RF part LL-450-553, **davisrf.com**.

[5]Palomar Engineers FB-56-43 ferrite beads, **palomar-engineers.com/ferrite-beads**.

[6]Several versions of *EZNEC* antenna modeling software are available from developer Roy Lewallen, W7EL, at-**eznec.com**.

Joel R. Hallas, W1ZR, is a *QST* Contributing Editor. You can reach him at **w1zr@arrl.org**.

# Easy Super Stealth Antenna

*A random-wire antenna that's easy to put together — and easy to hide if you live with an HOA.*

Dennis Lazar, W4DNN

**Figure 1** — My antenna arrangement. I do not live in a deed-restricted community so the unun does not need to be invisible.

At a meeting of our local radio club, Peter, K4PJG, a newly minted General-class licensee, was sharing his struggles living in a deed-restricted community with a homeowners association (HOA) that prohibits antennas of any kind. He'd only been able to operate portable while on trips in his camper.

Dave, K4DES, suggested that he might run a thin wire from the side of his home to a nearby tree, but even that would be too visible for the strict HOA rules. Always up for a challenge, I volunteered to put together a truly invisible antenna for Peter.

## Design Plan

Like my own roof, Peter's uses barrel tiles. Unlike asphalt shingles, the tiles stand off the roof structure by about 3 inches, are non-conductive, and do not absorb moisture. The ideal invisible antenna, I thought, would be an end-fed random wire in the same color as the roof. Simply draped across the roof, it should blend in with the texture and complex pattern of the tiles, rendering it virtually invisible. But I was unsure about what kind of RF pattern this antenna would produce. I had read that very low wire antennas might have omni-directional patterns and not be entirely NVIS (near-vertical incidence skywave).

At the time, I was also interested in a second antenna for my own home station, so I used my home to experiment with setups to help Peter. Our homes, like many in Florida, are constructed of concrete block covered by stucco. The outer stucco finish is a type of plaster secured to the outside surface of the block by a fine aluminum grid. This results in the house becoming a perfect Faraday box. RF has a very difficult time getting in or out. Therefore, there is no reception. The new antenna had to be outside the structure and low enough not to be a good conductor for lightning.

## Construction

In the garage, I had a 40-meter dipole, but due to a 2004 hurricane, I no longer had trees.

**Figure 2** — Peter, K4PJG, prepares to deploy the invisible antenna wire, which is secured at one end to the TV dish.

So, I repurposed the black-jacketed 14-gauge stranded wire by simply draping 44 feet of it across the roof, securing it at each end with black Dacron cord and attaching a length of RG-58U coax long enough to reach the shack (see **Figure 1**).

To arrive at the optimal wire length for a random wire antenna for the ham bands, I assessed a range of wire lengths to fit the available space. Any length chosen must be far from being resonant on any of the ham bands. By using a random, non-resonant wire and a tuner, the antenna will be multiband.

Once the wire was on the roof, I listened to 20 meters. As a receiving antenna — despite the tremendous impedance mismatch — reception was fantastic. Now I needed to transmit. This would require a 9:1 unun (an impedance transformer from unbalanced line to unbalanced antenna), which would probably be easy to build, but I elected to buy one.

The final arrangement is simple. The unun is tied to my TV antenna tower with some Dacron cord. It attaches to the end of the wire, which is draped across the roof. I added a 30-foot length of counterpoise wire that runs down the side of the house and along the concrete footer parallel with the roof and antenna. It is positioned on the concrete so as not to touch actual ground. I clamped five ferrites around the coax where it hangs below the roof. This and the counterpoise wire keep the antenna from using the coax shield as a counterpoise, which would bring RF into the house.

## Testing the Setup

Once deployed, I fired up the rig in the shack on 20 meters and prepared to adjust my antenna tuner for minimum SWR. Before starting to tune, I called CQ at 5 W. Immediately, Pat, VE3DZZ, in Ontario, came back and gave me a 439. A few minutes later, after achieving a match at 1.2:1, and upping the power to 90 W, I made contacts across the US and as far as Cuba and Croatia. The antenna did not seem to be very directional.

To conduct a more objective test, I used the Reverse Beacon Network to monitor the bands and generate a list of the stations that have heard me. I called VVV de W4DNN six times using my vertical, with 14 replies. Using the rooftop wire, I received reports from 11 stations. Each station provided the signal-to-noise ratio of the received signal. The test results showed that most stations who heard me with the vertical also heard me with the wire, but at reduced strength.

Responding stations were located in

Washington, Texas, Kansas, Wisconsin, Ohio, Massachusetts, New Hampshire, Arizona, Colorado, North Carolina, Calgary, and Quebec. The only responding DX station was from Belize, and only the wire antenna was heard. The band was not open into Europe.

## Applying the Research

With my research finished, we got to work on Peter's stealth antenna. Because Peter's tile roof is white, he acquired white wire, white Dacron line, and a white 9:1 unun.

We positioned the unun near the shack window, lying flat on the roof and draped the wire across, toward the far side of the house, securing it with Dacron line. From the road and from any point around the house, the antenna is totally invisible, even up close (see **Figure 2**). Because one end of the wire is secured to the satellite TV antenna, it appears to be part of the TV installation.

## On The Air

Once deployed, it was time to get the new antenna on the air. Angelo, K3YUN, used his antenna analyzer to tune the wire to a perfect 1:1 match on 20, 40, and 80 meters. A few days later, a great contest weekend provided several perfect opportunities to work many stations in the US and DX as well (see **Figure 3**).

In the weeks following, Peter has been working the world, mainly on 20 and 40 meters, with the neighbors blissfully unaware that there is now a ham operating in the neighborhood.

**Figure 3 — Following tune-up, Angelo, K3YUN, called the first CQ with Peter anxiously observing.**

---

All photos by the author.

Dennis Lazar, W4DNN, has been licensed since 1960. After leaving the Coast Guard, he worked with NASA on the Apollo moon exploration project. Turning to journalism, he was associate editor of two trade magazines and later, editor and publisher of a weekly neighborhood newspaper. Returning to college, he became a registered nurse therapist and then doctor of naturopathy, working in a private pain-management practice until 2008. Dennis holds an Amateur Extra-class license and a general commercial license. He has served as QRP editor for *CQ* Magazine, and has written many articles for *73*, *CQ*, and *QST* magazines.